JAMES LICK'S MONUMENT

James Lick. (Reproduced by kind permission of the Mary Lea Shane Archives of the Lick Observatory.)

James Lick's monument

The saga of Captain Richard Floyd
and the building of the Lick Observatory

HELEN WRIGHT

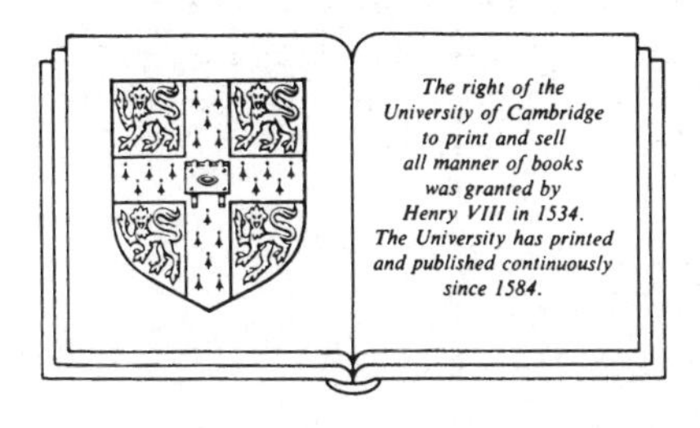

CAMBRIDGE UNIVERSITY PRESS

Cambridge

London New York New Rochelle

Melbourne Sydney

PUBLISHED BY THE PRESS SYNDICATE OF THE UNIVERSITY OF CAMBRIDGE
The Pitt Building, Trumpington Street, Cambridge, United Kingdom

CAMBRIDGE UNIVERSITY PRESS
The Edinburgh Building, Cambridge CB2 2RU, UK
40 West 20th Street, New York NY 10011–4211, USA
477 Williamstown Road, Port Melbourne, VIC 3207, Australia
Ruiz de Alarcón 13, 28014 Madrid, Spain
Dock House, The Waterfront, Cape Town 8001, South Africa

http://www.cambridge.org

First published 1987
First paperback edition 2003

A catalogue record for this book is available from the British Library

Library of Congress Cataloguing-in-Publication Data
Wright, Helen, 1914–
James Lick's monument.
Bibliography: p.
Includes index.
1. Lick Observatory – History. 2. Lick, James,
1796 – 1876. 3. Floyd, Richard S. 4. Astronomers –
United States – Biography. I. Title.
QB82.U62M689 1987 522′.19794′73 86 – 4179

ISBN 0 521 32105 0 hardback
ISBN 0 521 53455 0 paperback

Contents

Illustrations

Preface

Many years ago I went to the Lick Observatory on Mount Hamilton in California in search of material on the noted astronomer George Ellery Hale. There I first learned of Captain Richard S. Floyd, president of the Lick Trust, who, with Thomas E. Fraser, was largely responsible for building James Lick's observatory. Mary Lea Shane, wife of Donald Shane, then director of the Observatory, and I began to delve into the early records of the Lick Trust. We found these records in a safe behind the director's office, where they had been since the completion of the Observatory in 1886. They had been pasted into old scrapbooks, making them hard to file. To Mary's initial horror I suggested ripping the correspondence out of the scrapbooks and arranging it according to author or subject. Soon the floor of the director's house was covered with piles of handwritten letters that told the vivid story of the founding of the Observatory by the millionaire James Lick, who had arrived in San Francisco in 1848 and, while thousands rushed off in search of gold, had invested in land.

At the same time, in search of material from a later period, I hauled out huge redwood boxes stored under the great 36-inch refractor dome and soon realized what a wealth of astronomical history they contained.

The quest for the story of the founding of the Lick Observatory and of Floyd's role in building it has taken me far afield, from California to Louisiana, Georgia, and Washington, D.C. During the building of the Observatory, Cora Matthews, niece of Cora

Lyons Floyd, Floyd's wife, stayed with the Floyds on Mount Hamilton. She wrote often to her relatives in St. Francisville, Louisiana. These letters, one of the few sources of personal material on that important period, fell into the hands of a friend who stored them in her attic until fire destroyed the house and everything in it.

After Floyd's death, most of his family papers, including portraits, diaries, and letters stored at Kono Tayee, Floyd's estate on Clear Lake, California, were destroyed by people who had no sense of their value to the Floyd family and posterity. The mansion suffered the same fate.

One of the hardest aspects of biographical writing is the accurate tracing of history through a web of misrepresentation. This, unfortunately, was the case with Edward S. Holden, who was to become director of the Observatory on its completion in 1888. Not content with some credit for building it, he wanted all the credit, and as the years passed, this desire grew. A prolific if egocentric writer, he wrote countless articles in which, as the "chief adviser to the Lick Trust," he claimed responsibility for the entire project. As a result, some modern historians, relying on his published books and papers, and a biased view of the primary source material in the Mary Lea Shane Archives of the Lick Observatory and elsewhere, have given him the glory he craved and have ignored the true story of the building of the Observatory.

Luckily, the files of the real "chief adviser" to the Trust, Simon Newcomb, at the Lick Observatory and in the Library of Congress, have not been destroyed. Luckily too, early staff members at the Observatory realized how Holden misrepresented the facts. A former director, W. H. Wright, wrote of Floyd, "He brought to the task a resourcefulness bred of the exigencies of pioneer life. That this quality was possessed to an unusual degree by Captain Floyd no one who has had the advantage of reading firsthand accounts of the Observatory's history can possibly doubt." And again: "To his ability, resourcefulness and industry, the success of the whole observatory project, is largely due." As Mary Shane put it, "In retrospect one may well wonder how the Observatory would have fared without Captain Floyd's dedicated leadership."

In that pioneering task, Floyd was fortunate to have the help of the high-spirited Thomas E. Fraser as superintendent of construction. Of Fraser, Floyd wrote, "To his unceasing care, great technical knowledge, and real comprehension of purely astronomical ideas, the excellency of the observatory is due, in no slight degree." Fraser's letters and his "log" kept all through the building of the Observatory are a vital part of this saga. They are characterized by his unique spelling. Floyd, too, wrote in his own individualistic way, with a spelling that was often nonconformist. These spellings have been retained in quoting the two men in this book.

Acknowledgments

The greatest satisfaction of biographical writing is the friends an author makes along the way. This has been particularly true in my quest for the story of Captain Floyd and the building of the Lick Observatory.

My thanks go first to Donald and Mary Shane. Astronomers the world over had no better friends. Mary, who shared my interest in the early history of the observatory, had hoped to write that history herself and had made a beginning on the project. She died in 1983, on her eighty-sixth birthday, without having realized that hope.

She would have been glad to know that, in 1984, Cambridge University Press published Donald Osterbrock's excellent biography of James Keeler, the second director of the Lick Observatory. With many other historians, Dr. Osterbrock profited greatly from the years that Mrs. Shane, with able volunteer help, devoted to organizing the Lick archives. I hope that she would also have liked my account.

Through the Lick archives I made other friends who helped me along the way. It gives me great pleasure to thank Donald Osterbrock and his wife, Irene, one of Mary Shane's volunteers, who became an expert on the clippings in the Lick files. Dorothy Schaumberg, now curator of the Mary Lea Shane Archives, has carried on the work begun by Mrs. Shane, adding important collections to the Lick archives. I am especially grateful for her help in obtaining the Lick archival photographs and arranging for their reproduction by the Lick Photographic Laboratory. In that laboratory, Diana Larson deserves my special thanks for her efforts to obtain the fine reproduction of the *Lady Adams*, the ship on which James Lick arrived in San Francisco.

Another good friend made through the Lick archives is Jeffrey Crelinsten, a Canadian historian of astronomy, who found vital material for his thesis on astronomers and relativity there.

The writing of this book has been helped by many other people. I would like to thank all those who have provided the source material listed in the notes and in the captions for the illustrations. These include the manuscript divisions of the Library of Congress and the New York Public Library, the Smithsonian Institution Archives (especially Susan W. Glenn, who has been most helpful in the examination of the Joseph Henry Papers), the Society of California Pioneers, and The United States National Archives and Records Service. My thanks go to James D. Hart, director of the Bancroft Library of the University of California at Berkeley, and Bonnie Hardwick, head of the Manuscript Division there, for the valuable materials described in the notes. To Daniel H. Woodward, librarian at the Henry E. Huntington Library and Art Gallery, for permission to use the watercolor of the *Lady Adams* from John Hovey's *Journal of a Voyage (Around the Horn)*. At the California Historical Society in San Francisco, my thanks go first to Stephen J. Fletcher, assistant curator of photographs, and to Douglas Haller, curator of photographs, who gave me permission to use the photographs from its fine collection.

In Lake County, California, the county historian, Marian Geoble, was generous with her broad knowledge of that county's history.

In St. Marys, Georgia, Eloise Bailey, who has long been interested in the history of the Floyd family, helped me greatly. My thanks to her and to Picot Floyd, who gave me permission to examine the Floyd papers, which he deposited at the Georgia Historical Society in Savannah.

In St. Francisville, Louisiana, Elisabeth Kilbourne Dart encouraged me all along the way. In Baton Rouge, at Louisiana State University, I thank the archivists for access to the papers of Henry Lyons and the Bowman family, which provide a picture of antebellum plantation life.

In Washington, D.C., at the National Archives, important material on Floyd's days in the Confederate Navy was found. At the former Naval Observatory in Washington, D.C., Jan K. Herman, historian, took us on a fascinating tour of the old building, now under the direction of the Naval Medical Command. Unfortunately it is in disrepair, but it is hoped that funds can be found to save it. At the present Naval Observatory, also in Washington, librarian Brenda Corbin was most helpful. I thank her for permission to use the excellent copy of Simon Newcomb's portrait, and the lithograph of the old observatory given to me by Jan Herman. This was made around 1876, when President Floyd called there to see Simon Newcomb, chief adviser to the Lick Trust. To David DeVorkin, a well-known astronomical historian, I am grateful for the photograph of Howard Grubb taken in 1877. In the Manuscript Division of the New York Public Library, the Henry Draper papers have been useful. Francis McAdoo gave me permission to use the W. G. McAdoo papers at the Library of Congress in Washington, D.C. My thanks to him and to Brice Clagett, who provided me with part of W. G. McAdoo's diary.

Some of the greatest contributions to this book came from those who read the manuscript and offered valuable suggestions and criticism. One of the first to read it was my cousin Winifred Heron, wife of David Heron, then Director of the University of California at Santa Cruz Library, and one of Mary Shane's volunteers in the Lick archives. Her constructive suggestions and encouragement have been most helpful. Another reader was Katherine Gordon Kron, a fellow astronomy major at Vassar College. She was enthusiastic about the manuscript and urged its publication. Charlotte Cassidy, an expert typist, copied the first draft and promoted the publication of the book from the beginning. In Ottawa, Canada, Jeffrey Crelinsten, an astronomical historian and scientific administrator, read the draft and the revised manuscript and made valuable comments based on his broad knowledge of the history of astronomy.

Another helpful reader was Eloise Bailey of St. Marys, Georgia. In St. Francisville, Louisiana, Stephen and Elisabeth Kilbourne Dart read the original draft and encouraged me in countless ways. In Montreal, Canada, Susan Weston Smith read the manuscript from a layman's point of view and made useful suggestions. Edward J. Pershey, the historian of Warner and Swasey, took time from his busy schedule to read the manuscript. I thank him. In Santa Cruz, at the Lick Observatory archives, Dorothy Schaumberg read the manuscript from her expert point of view and approved it. Don Osterbrock read the manuscript with a critical eye, concluded that I had really written two books instead of one, and proposed that I drop most of the Floyd background material and concentrate on the story of the founding of the Lick Observatory and Floyd's role in that building. This view was shared by David DeVorkin. Cambridge University Press accepted this view and agreed to publish the present book.

I would also like to thank Barbara Giffen for providing me with shelter during the many weeks I spent in Santa Cruz.

For additional help in typing the manuscript, I thank my brother William Finley Wright, Lawrie Chase, Stirling Wright, and Michelle Lucarelli. For her excellent work on the index I thank Joan Bothell.

In 1988 the Lick Observatory will be one hundred years old. Although a great amount of material has been lost through fire and other forms of destruction, it is lucky that so much has been saved to give us a vivid picture of the founding of the Observatory and the revolutionary achievements of astronomers who have built and used its telescopes.

Prologue

The Great Comet of 1843 flared across the sky, rousing wonder in the hearts of the people and fear in those who saw in it the sign of the world's end. On September 8 in that same year Richard Samuel Floyd was born at Bellevue plantation in Camden County, Georgia to Charles Rinaldo and Julia Boog Floyd. He was born to a star-studded destiny no one could have foretold.

In 1843 James Lick, a Pennsylvania Dutchman, was living in Lima, Peru, making frames of beautiful polished woods for the finest pianos in the city and hoarding the pile of gold he would carry to California just seventeen days before the discovery in Sutter's Creek. Yet he was to gain his fortune not in gold but in land. This land would give him the power and the wealth to found the Lick Observatory on Mount Hamilton above San Jose in California. This, in turn, would lead to a fortuitous meeting with Captain Richard S. Floyd and to Floyd's appointment as president of the Lick Trust, in charge of building the greatest observatory with the largest refracting telescope in the world.

What happened in those years after the comet to make the eccentric Lick entrust his dream to the thirty-two-year-old Floyd, who was forty-six years his junior? What forces in Floyd's background and training made him equal to the task? These are the threads we shall try to unravel in telling this story.

1

The story of James Lick

This, as Floyd heard it from Thomas Fraser, the foreman on Lick's ranch, was the story of the millionaire's life. Born August 25, 1796, in Fredericksburg (then Stumptown) in Lebanon County, Pennsylvania, Lick learned the craft of cabinetmaking from his father, John Lick. Later, in Baltimore, he applied his skills to piano making. Fiery and impulsive, with a taste for adventure, he sailed from New York to Buenos Aires on April 1, 1825. He worked there for several years, then set out for Valparaiso, Chile, around stormy Cape Horn. In 1836, after three years in Chile, he moved on to Lima, Peru. Along the way he made and sold piano cases, accumulating a small fortune in silver coin that, when converted into Peruvian gold doubloons, amounted to $30,000. In Lima, in addition to cutting and working brass for his pianos, he ran a theater and an amphitheater for bullfighting and dabbled in mercantile ventures.[1]

In 1846, learning of the Mexican War and the occupation of California by the United States, he decided to go there. A friend tried to dissuade him from risking his life among "cut-throats and ruffians" in a strange country with an uncertain future. But Lick said he had always been able to take care of himself in the face of any obstacle. He had noticed, too, that once the United States got hold of a territory, it was not apt to let go.[2] With his hoard of gold, six hundred pounds of Guatemalan chocolate made by his friend Domingo Ghirardelli, his workbench and tools, he boarded the *Lady Adams* and sailed for San Francisco

The *Lady Adams* of Baltimore. (Reproduced by kind permission of the Henry E. Huntington Library and Art Gallery.)

in 1847.[3] He arrived there on January 7, 1848, just seventeen days before James Marshall discovered a gold nugget at Coloma near the future site of Sacramento. Landing in San Francisco, Lick found a noisy village of eight hundred inhabitants, many living in tents or shanties. Its few houses, lit by oil lamps and candles, stood on sandy, often muddy streets, littered with bottles discarded from the windows above. It was a dreary sight.[4] Yet, while hundreds rushed off to the gold fields, Lick, arrayed in a long overcoat and tall plug hat, made one trip only to the diggings at Mormon Island on the American River.[5] He did not like it and hurried back to San Francisco. There he envisioned "the waterfront noisy with the turmoil of trade." [6] He started to buy land in and around San Francisco, including many town lots. Soon he extended his holdings to the region of San Jose in Santa Clara County, then the state's capital.

In 1852, three miles south of Alviso on Guadalupe Creek, three miles north of Santa Clara, and five miles northwest of San Jose, Lick bought and started to rebuild a huge mahogany gristmill at a cost of $250,000.[7] When it was finished he considered it second to none in elegance. For it he ordered from Boston the finest machinery, of glittering brass and shining steel, "as finished as that of a present day steamboat." It was soon turning out the "finest flour in the West." Lick lived in a little frame shanty nearby and slept on a mattress atop a grand piano.

Strong and wiry, with great powers of endurance, Lick worked from dawn until sundown, cultivating his beloved flowers and exotic fruit trees that flourished in this fertile valley. He ordered rare plants and shrubs from many parts of the world. Along with trees from Australia he transported shiploads of their native earth. Some who saw him picking up old bones thought him crazy, but ground up, such bones made excellent fertilizer. He became, in fact, a pioneer horticulturist.

In 1860 Lick decided to build a colonial mansion near the mill at Agnew.[8] Like the mill, it was made of the finest materials. It had twenty-four rooms. Yet he occupied only the living room – with two or three chairs and a bookcase filled with books on metaphysical and scientific subjects. On top of his piano in a corner he had a roll of blankets. He had no carpets and no curtains. Instead there were rows and rows of newspapers, with fruits of all kinds drying on them. Lick spent $250,000 on the house and its beautiful grounds. "He lived among his flowers and wondered about the universe. It fascinated him, the mystery of it all."

But each winter, when the rains came, the Guadalupe overflowed its banks and ravaged his beautiful orchards. In 1870 Lick abandoned that house.[9] He owned a 105-acre tract in San Jose, fronting on First Street and running out to the city limits. There he built the Lick homestead with three large hothouses. Soon his neighbors saw cart after cart going in a caravan from the mill to the homestead as Lick moved all his trees and plants to his new property. Lick himself, "looking like a tramp picked up on the road," rode on a board in one of the carts next to one of his laborers, overseeing the work. For two years this "parade of swaying trees and plants" went slowly over the roads.[10] Among his trees were two locusts he could not transplant. He cut them off at the base and sawed the wood into veneering for pianos. Lick told his friend Conrad Meyer that he had used a similar, though softer, form of locust for a few octaves in the base of one of his pianos in Peru. "It did make the most powerful and also the sweetest tone I ever did hear, without exception, in all my long experience in pianos."[11] He asked Meyer to use this to make a piano for his parlor at the Lick House.

A friend who knew him at this time wrote, "Lick was a tempestuous old fellow in his dealings with humankind. He often found them ridiculous. Their stupidity irritated him. But with his flowers he was gentle and tender. His days were spent among them, hovering over the plants as if they were capable of love. The flowers bloomed luxuriantly, as if to thank him."[12]

As the years passed, Lick's legendary reputation grew. Reports of his unorthodox characteristics and curious habits spread. A lean-faced man, with sharp blue eyes, thin features, and a tufted beard that lined his jaw, he wore "ill-sorted clothes flung carelessly on him, as if time spent in dressing was time wasted."[13] Thus a contemporary portrait. Others, commenting on his retiring disposition, said he tended to be thoughtful, often moody. In bargaining he was quick-witted and longheaded and hated above all to be overreached.

Yet it was this recluse who late in 1861 began building the Lick House, a magnificent three-story red-brick building at the corner of Sutter and Montgomery streets in San Francisco's heart. Opened in 1862, it soon became the leading hostelry in San Francisco, the gathering place for the city's elite. Its mahogany lobby was decorated with red plush sofas and chairs. Its high-domed dining room, with flagged marble floors, was modeled after that at Versailles.[14] Lick himself finished its parquetry of rare woods from South America and the Orient, as well as its polished rosewood picture and mirror frames. On its walls he hung paintings by William Keith and Thomas Hill – one a spectacular view of a ship coming through the Golden Gate in

1849. Here, in the restaurant known as the Palace of Fine Arts, a fabulous "Free Lunch," served with drinks from 4:00 to 11:00 P.M., included such delicacies as Bolinas Bay clams, smoked salmon, Holland herring, and terrapin stew.[15]

Lick's appearance contrasted sharply with that of his elegantly dressed guests. A guest describes his first view of the proprietor, in the "oddest looking turnout imaginable," making his way through the crowd of carriages to the front entrance. His horse was old and weary; his buggy, covered with dust, dilapidated, and broken, was held together with baling wire and contained a roll of blankets. He was poorly dressed and apparently ill-fed. Yet he did not look like a mendicant. "There was no humiliation or self depreciation in his countenance. His eye even had a satisfied, self-possessed expression. . . . As he alighted, a dozen men rushed out to greet him and shake hands." [16]

In time Lick extended his holdings to the south. On June 6, 1865, he bought for $5,000 the 6,647-acre Los Felis Rancho, a property he was to own until his death.[17] If he could have returned a hundred years later, he would have been amazed to find himself on Los Feliz Boulevard, driving through the heavily populated region of North Hollywood and Glendale where this, one of the loveliest ranches in California, had once stood. In a complex deal soon afterward, he also became the owner of beautiful Santa Catalina Island – 45,820 acres.[18]

From the Lick House and all his other properties, Lick, in twenty years, accumulated some $3.5 million. Then, like many aging millionaires, he wondered what to do with his wealth. By 1873, various plans had begun to take shape in his mind. One that caught his imagination was the building of the largest telescope in the world.

Many have since asked, "Why did Lick dream of such a telescope?" No one is really sure. Some claim he had a mystical belief in the cosmos as God's heavenly handiwork. Others tell of a priest in Rio de Janeiro who, feeling sorry for the solitary man, talked with him of ancient beliefs about the heavens in such a way that Lick decided then that, if he ever had the power, he would do something to add to the understanding of the universe.[19]

George Schönewald, who had leased and managed Sam Brannan's fabulous hotel at Calistoga Hot Springs before becoming manager and maître d' at the Lick House, told of a day he and Lick were walking down Montgomery Street. Suddenly Lick stopped to exclaim, "One day men will go to the moon and back."[20] Schönewald thought him crazy but said nothing.

From the daughter of one of Lick's few friends comes another tale. Lick, who always carried a big blue silk handkerchief with great stars on it, often served wine to her father and exchanged books with him. Those he liked best were nearly always about philosophy or science. "Gold coins held no charm for him," she said, "His books and his flowers held all his love. A pet book was one that showed the Egyptian religion to be founded on astronomy."[21] Lick was devoted also to the works of Andrew Jackson Davis, the American spiritualist. Like Poe's *Eureka,* Davis's works give a poetic view of the cosmos.[22]

As men of different sorts claimed responsibility for persuading Lick to build a large telescope, other stories emerged. One of the earliest was a Portuguese American, George Madeira, student of astronomy and geology, who, in 1852, at the age of fifteen, had crossed the continent carrying his star charts. In 1860, he wrote later, he was lecturing in San Jose when a gentleman stepped up and asked, "Will you accept an invitation to visit my place and remain for a few days?" Madeira, surprised, accepted. Each night he, with his host, James Lick, gazed at the skies through his little telescope.

A few years later, according to Madeira, the two men met again. Madeira had mounted a 6-inch refractor near Volcano in Amador County in the Sierras. Again he invited Lick to look through his telescope. He talked of the big telescopes of Lord Rosse and Sir William Herschel, and their discoveries, and then exclaimed, "If I had your wealth, Mr. Lick, I would construct the largest telescope possible to construct." Madeira subsequently felt that he had planted the first vital seed in Lick's mind.[23]

It is likely also that two of the leading scientists of the day helped to persuade Lick to use his wealth to cultivate science for the benefit of mankind. One was Joseph Henry, the august secretary of the Smithsonian Institution, in Washington, who in 1871 happened to meet Lick at the Lick House.[24] When Lick asked what he should do with his millions, Henry may well have told him of the Englishman James Smithson and his unusual bequest to a country he had never seen for the founding of an institution for the "increase and diffusion of knowledge among men." (Out of this bequest, the Smithsonian Institution was born.)

The other was the famous Swiss naturalist Louis Agassiz, who rounded Cape Horn and landed in San Francisco with the Hassler expedition in August 1872. During the month he spent there, the long-haired scientist lectured at the California Academy of Sciences. He told of his research and urged the members to support science. He looked around at the academy's cramped and dingy quarters, called its empty treasury a reproach to the young state of California, and said, "I hold it one of the duties of those who have the means, to help those who have only their head, and who go to work with an empty pocket."[25] A few months later, in February 1873, Lick, doubtless moved by Agassiz's appeal, announced the gift to the academy of a valuable lot on Market Street for its headquarters.

The following year Lick was asked to accept the presidency of the Society of California Pioneers, to which he had given a valuable lot on Montgomery Street in 1859. He hesitated but then agreed, on condition that his friend David Jackson Staples serve in his place "under any and all circumstances."[26] Several visits between the two men followed. One day Lick asked Staples to witness the signing of his will. Staples, impressed by Lick's wisdom, could hardly suppress a smile, because some of his ideas were so odd. They ranged from a marble pyramid, larger than that at Cheops, to be mounted on the shore of San Francisco Bay, to giant statues of Lick's father and mother and himself, to be executed by the world's greatest masters and erected on North Beach, where he owned extensive property. On his Market Street land, in San Francisco's heart, he envisioned a telescope larger and more powerful than any yet built. For the telescope he allotted $500,000; to his illegitimate son, the paltry sum of $3,000.

After reading this extraordinary document, Staples protested, "Mr. Lick, I am not an astronomer, but I know well enough Market Street is not the place for an observatory. In the first place, the climate is too foggy, and, in the second place, the traffic on the streets would disturb the instruments." [27]

As to the treatment of his son, Staples spoke bluntly: "Everybody believes he is your son. You know you have got him by that Dutch woman." Instead of the "miserable" sum proposed, Staples urged a handsome fortune. Like everyone else, he had heard the tale of John's birth to a rich miller's daughter whom Lick had loved in his youth.[28] When James asked permission to marry the beautiful Barbara, the miller refused to have anything to do with the young suitor, taunting him about his poverty and lack of education. Young Lick swore then that one day he would own a mill finer than anyone in his native town had ever seen. He left home and never again saw his Barbara. But, near Alviso, as we have seen, he rebuilt the fantastic mahogany mill that became known as "Lick's Folly."

It was at this mill that his illegitimate son John, a pale, weak-faced young man, appeared in 1855. Lick took him in, but they were so different that the relationship was never a happy one. John was slow-moving, irritable, without initiative. He had no interest in the flowers his father cultivated or in the books he loved. He did as he pleased. In 1863 he returned home, then came back to stay with his father until 1871.[29]

Now, in answer to Staples's tirade, Lick tried to explain his will. One day, when he was away from home, he had left his beautiful parrot in his son's charge; on his return he found the bird neglected, the cage dirty. He could never forgive such negligence. He was damned if he was going to leave more than $3,000 to a fellow of that sort. Staples argued that no court would accept such a will and got up to leave. "Well, good morning," he said. "Of course you do not want to see me any more." Lick, beside himself, called his friend back, demanding, "Well, what shall we do with the money? You know everybody," he cried. "Go out and consult your friends!" [30]

Staples, moved by Lick's concern, hurried off to see some leading San Franciscans – Governor Newton Booth, Chief Justice William Wallace, and William C. Ralston. At Ralston's hilltop home they met with Mayor Thomas Selby, Darius Ogden Mills, the banker, and Daniel C. Gilman, president of the nascent University of California.[31] Staples told them of Lick's desire to dispose of his estate in a way that might do the greatest good, yet with the greatest possible credit reflected on him by his beneficiaries.

All agreed that Lick must be persuaded to abandon his dream of monumental statues for more worthwhile causes. They devised various wills, and when Staples presented them, he advised Lick against his dream of statues that would become valuable antiques: "Very likely we shall get into a war with Russia or somebody and they will come around here with warships and smash the statues to pieces in bombarding the city." [32] Thus the observatory took the place of the pyramid. "The beauty of the one was to find a substitute in the scientific use of the other. The instruments were to be so large that new and striking discoveries were to follow inevitably, and, if possible, living beings on the surface of the moon were to be described, as a beginning." [33]

Meanwhile George Davidson, president of the California Academy of Sciences, to which Lick had donated land on Market Street, arrived to thank him and promote his scheme for a great telescope. Davidson, an Englishman of Scottish descent, had come to the United States as a child. In charge of charting the Pacific coastal waters for the U.S. Coast Survey, he had ranged from Lower California to Alaska under rugged conditions, in all kinds of weather. In California he climbed the High Sierras and dreamed of mounting a large telescope on one of those peaks. In 1872 he published a report entitled "Astronomical Observations at Great Elevations in the Sierra Nevada at 7200 feet." [34]

In 1869 Davidson had written to Benjamin Peirce, then head of the Coast Survey, "I have been feeling the ground in this community, among men likely to be interested, about the establishment of a large observatory on this coast, with the completest outfit of instruments, observers, means of publication etc."[35]

Now, meeting with Lick, Davidson did all he could to encourage the old man's interest in astronomy, talking of Herschel and Rosse, just as Madeira had done. It was even said that Davidson arranged for a small telescope through which Lick could gaze at the skies through his bedroom window. Lick's mind was doubtless prepared by the diverse seeds already planted there. But Davidson, who noted that Lick had a strong mechanical bent, always felt responsible for persuading him to give a large share of his wealth for a great telescope, to be mounted at 10,000 feet in the Sierras. It was certainly Davidson who announced Lick's intentions to a large audience at the California Academy of Sciences on October 21, 1873. After lecturing on spectrum analysis, he told his listeners how impressed he was with Lick's comprehension of original research in cosmic physics and how enthusiastic he was about the plan to build a telescope larger and more powerful than any yet made. He concluded, "A thousand years hence, the James Lick Observatory, endowed with perpetual youth, will continue to unfold the mysteries of the cosmos, and to search for new worlds to conquer."[36] In bold headlines the *Daily Alta California* proclaimed, "THE LICK OBSERVATORY, THE GREATEST SCIENTIFIC WORK OF AMERICA ABOUT TO BE COMMENCED."

On April 1, 1873, Lick had returned alone to his house at night. Suddenly he became dizzy and collapsed on the floor, suffering from a paralytic stroke. There his ranch foreman, Tom Fraser, found him the next morning, struggling to get up. He took Lick in his horse-drawn buggy to a doctor, who assured the worried Fraser that with rest and care, Lick would be able to use his arms and legs again. Fraser took him home and put him to bed. The following day he asked Lick if, instead of staying at the homestead, he would prefer to go to the Lick House. Lick nodded. There, in a small room, he recovered slowly, planning for the distribution of his wealth. From this time on, Lick relied increasingly on Fraser's advice. "Until the day of his death no man was closer to Lick than Captain Fraser."[37]

Thomas E. Fraser. (Reproduced by kind permission of the Mary Lea Shane Archives of the Lick Observatory.)

A skilled carpenter and fine artisan, Tom Fraser had been a ship's cabinetmaker in his native Nova Scotia. With his brothers, John and Howard, he arrived in San Francisco in 1866. Soon afterward he found work at the White Copper mines in Nevada. When trouble erupted there, he joined his brother Jack at the

Sulphur Bank mines at Clear Lake in northern California. There, in 1869, he first met Dick Floyd. It proved a lucky meeting. Tom was tall and swarthy, sported a mustache, and spoke with a trace of a Scottish burr. Like his brother he inherited the Nova Scotian's ability to do anything with his hands. He hated idleness. According to Jack's son, if his father asked him to get a hammer he did not walk – he ran. He would not have dared stand with his hands in his pockets.[38] A hardy Scot, Tom was, like Jack, "well adapted by nature for the hardships of pioneer life, coming of stock whose tastes led them to exploration and blazing the way for the less venturesome."[39] While working as a contractor in San Francisco, Fraser had first met Lick, a lover of gardens and orchards, who had ordered replicas of the iron, wood, and glass conservatories at Kew Gardens outside London. Built by Lord and Burnham in Irvington, New York, they were shipped in large crates around the Horn. Lick wanted Fraser to assemble them at his homestead near San Jose, but this would never be done. After Lick's death, they would be mounted in the conservatory in Golden Gate Park.[40] But Lick, impressed by Fraser's "quick intelligence, prompt obedience and business sagacity," soon gave him other orders, and in 1873 he asked him to become his confidential agent and foreman.

Seven months later, just two days after Davidson's dramatic announcement of Lick's gift to the California Academy of Sciences, Lick, in a letter to Joseph Henry, told of his plan to construct an observatory to "rank first in the world." He asked Henry's advice on the building of the telescope, "as near a perfect instrument as the best scientific knowledge and skill and human workmanship can make, and that shall surpass in power anything yet attempted."[41]

Henry, trying to educate the visionary Lick, described the two types of observatories – the one devoted to the measurement of positions and motions of the heavenly bodies; the other, to the study of physical phenomena of the heavens. He ended his long and detailed letter with the hope that Lick would live long and enjoy the consciousness of having, by his efforts, "obtained the power to advance human knowledge, and thereby a higher civilization."[42]

Months passed and Henry continued to worry about Lick and his observatory. On August 1, 1874, he had written to the great English scientist T. H. Huxley, "The public mind in this country is now directed to the importance of original scientific research and I think there is good reason to believe that some of the millionaires who have risen from poverty to wealth will in due time seek to perpetuate their names by founding establishments for the purpose in question."[43]

Henry then told of Lick's gift for an observatory to be the largest in the world and said he had just proposed to Lick that he appropriate immediately "say $75,000 for the purchase of instruments for an *AstroPhysical* observatory, and that he place this money in the hands of the best astronomical physicist who can be obtained, irrespective of country." By adopting this course, he noted, "the Observatory may almost immediately be put into operation and he will have a fair opportunity of witnessing fruit from it which will prominently connect his name with the history of science."

Lick, Henry knew, was eager to see his "monument" completed before he died. It was impossible to build the largest telescope in the world in a short time. He warned Lick, "On the other hand if you adopt the ordinary course of putting up an expensive building which, when finished, may be badly adapted to the uses for which it is intended, and if you wait until the largest telescope can be finished, your life will probably be terminated before any results are produced."[44]

But Lick, in 1874, was in no mood to accept any dilution of his grand scheme. Disregarding Henry's prediction on his life's end, his answer was brief. He thanked him for his suggestions and added, "At the present time however I am not prepared to take any steps in the matter, but when action is taken you may be assured I shall not be forgetful of your kind advice."

Henry had even suggested (August 4, 1874) that the noted British astronomer Joseph Norman Lockyer be appointed "Superintendent" of the new astrophysical observatory.[45] He referred Lick to a sketch of Lockyer in the November 1873 issue of the *Journal of Popular Science*. If Lockyer was offered "the opportunity of extending the bounds of human knowledge" through Lick's "enlightened liberality," Henry thought he would accept.

At the same time Henry warned Lockyer not to let this proposal interfere with any other plans, as Lick was said to be under the influence of another person, who himself wanted to be director of the establishment. This could only have been George Davidson.

On June 14, 1874, as Davidson was about to leave on a trip around the world, Lick told him he had added $200,000 to the original sum for the telescope, after learning that $500,000 would not be enough. Davidson had assured Simon Newcomb, the leading American astronomer, that the observatory proposed by Lick would be built. He said confidentially that $1 million in gold would be assigned to it.[46]

While a storm of controversy swirled around him, Lick went ahead with his plans for the telescope. After endless wrangling he issued his First Deed of Trust, "to be controlled and administered by a board of five of San Francisco's leading citizens." It was dated July 16, 1874. The trustees included Mayor Thomas H. Selby, who was in the smelting business, and Darius Ogden Mills, president of William C. Ralston's Bank of California, an owner of the California Borax Company and, by 1875, of the Sulphur Bank Quicksilver Mining Company at Clear Lake. A brilliant, conservative man many considered cold and austere, Mills had made a fortune in banking and real estate.

In the fall of 1874, Mills headed for Washington to consult with Newcomb at the Naval Observatory, where the "great" 26-inch refracting telescope recently had been completed by Alvan Clark and Sons. Apparently Lick had read accounts of this telescope in the newspapers that visitors had seen scattered around his room. It is likely that he had also read in *Scribner's Monthly* "The Story of a Telescope," in which Newcomb described the building of that telescope and the grinding of the lenses by Alvan Clark, the optical expert.[47] Lick's understanding of the real work of a telescope was limited. Said Newcomb, "I do not think that an observatory properly so called was, at first, in

Mr. Lick's mind; all he wanted was an immense telescope stuck up at the end of a long pole."[48]

Searching for an explanation of Lick's dream, Newcomb suggested that astronomy is a "science which seems to have the strongest hold on minds which are not intimately acquainted with its work. The view taken by such minds is not distracted by the technical details which trouble the investigator, and its great outlines are seen through an atmosphere of sentiment, which softens out the algebraic formulae with which the astronomer is concerned into those magnificent conceptions of creation which are the delight of all minds, trained or untrained."[49]

2

Captain Floyd and the Lick Trust

Lick's First Deed of Trust was dated July 16, 1874. On a memorable day late in that same year Richard Samuel Floyd, a native of Georgia, arrived at the Lick House. There Tom Fraser was waiting to introduce the exuberant, outgoing Floyd to James Lick, the lonely old millionaire who dreamed of building the largest telescope in the world.[1]

Lick was then becoming increasingly worried about the site of his telescope. On Market Street, where he had dreamed of mounting it, many people could see and admire it. At Lake Tahoe, the site investigated by the engineer A. W. von Schmidt, it would rarely be seen; in winter, access would be blocked by heavy snows. In the autumn of 1874 Lick consulted Fraser, who suggested instead the beautiful Mount St. Helena and said his friend Captain Floyd, who was "well versed in astronomy and had been in the habit of crossing the St. Helena mountain for a number of years," could supply further information. This led to the fateful meeting at the Lick House. They talked, and Floyd agreed that St. Helena, a flat-topped, extinct volcano, 4,345 feet high, clear of fogs, might be a good place to locate the observatory. He offered to do anything he could to help Lick with his project. "After an hour's converse," Fraser commented, "Mr. Lick, who was versed in human nature, was captivated. Floyd was a magnetic man, and charmed all with whom he came in contact."[2]

The more Lick learned about Dick Floyd, the better he liked

The Lick House, 1865. (Photograph by Addis and Koch, San Francisco. Reproduced by kind permission of the California Historical Society, San Francisco.)

what he learned. A cabinetmaker himself, he liked Floyd's account of his boyhood in Camden County, Georgia, where, in grandfather John Floyd's shipyard, he had built fishing and racing boats, developing the manual and inventive skills for which he had shown an early talent.[3] An ardent patriot, Lick was fascinated by Floyd's tales of his days at the Naval Academy in Annapolis, when on a two-month cruise to the Azores, Cadiz, and Madeira aboard the *Plymouth,* he learned many things that would prove unexpectedly useful years later. They included "every branch of seamanship," ranging from the use of the sextant and the taking of time sights by moon and stars, to the determination of latitude and longitude by the methods of Bowditch and Chauvenet. Yet Lick, with his pride in his grandfather Lick, who had fought under George Washington at Valley Forge, could understand Floyd's loyalty to the South in the Civil War. Floyd had entered the Academy in 1859 when he was just sixteen. On January 19, 1861, when Georgia seceded from the Union, he resigned to join the Confederate Navy. After serving on the *Huntress,* a Confederate steamer patrolling the Atlantic coast, Floyd was sent to the Bahamas to join the *Oreto,* later called the *Florida,* a commerce raider built in Great Britain and used by the Confederates to search out and destroy Yankee ships to prevent supplies from reaching Northern ports.[4] In the months to come the *Florida,* a "long, low, black rakish bark-rigged propellor ship," would range up and down the coast, from New England to Brazil, and across the Atlantic to neutral ports in France and England. It was a wanderer on the seas, without a home port, and its mission was dangerous, its actions daring. But the high-spirited Floyd reveled in that kind of danger. Their greatest prize was the *Jacob Bell,* a magnificent Yankee clipper homebound from Foochow, China, to New York, with a cargo of tea, firecrackers, camphor, and cassia, worth $1.5 million. The orders to the *Florida*'s officers were to destroy the ships they captured, after confiscating their cargoes, but to leave personal property untouched. The officers, crew, and passengers would then be released to a neutral ship or port. After all aboard the *Jacob Bell* had been transferred to the Danish barque *Morning Star,* Floyd and the crew set fire to the magnificent clipper, destroying her in a blaze made more spectacular by the Chinese fireworks.[5]

Later, in the neutral Brazilian port of Bahia, the *Florida* became the victim as it faced the Union cruiser *Wachusett,* under the command of Napoleon Collins. Collins, inflated by dreams of glory, decided to sink the *Florida* by ramming it head on and then escape by night.[6] Floyd, infuriated by this clandestine, unwarranted attack by an enemy ship in a neutral port, was shocked when, with his fellow officers, he was taken captive and imprisoned, first at Point Lookout, then at the old Capitol prison, and finally at Fort Warren, at the entrance to Boston harbor. When he was released he had to promise to leave the country immediately. With his good friend Tom Porter, he finally reached Liverpool in February 1865.

Meanwhile, as he would learn long after, the hapless *Florida* finally was sunk off Hampton Roads, Virginia. So ended the life of this valiant little cruiser that with its satellites, including the

Lapwing, Floyd had commanded. In all, sixty ships, valued at an estimated $4 million, were said to have been destroyed.[7]

On April 9, 1865, Robert E. Lee surrendered. Five days later Abraham Lincoln was assassinated. Floyd and Porter were still in England, still hoping for active service. As they considered the highly uncertain future, they could not decide where to go – Sonora, Paraguay, or China. They finally agreed to try their luck in Sonora, Mexico, then, "should circumstances favor, in California."[8]

On December 26, 1865, Floyd arrived in San Francisco, penniless and knowing no one. Willing to do almost anything, he later learned of the proposal of a new friend, who had arrived from Nova Scotia in the fall of 1866. John Fraser, born in 1844 and the youngest of thirteen children, had left his father's farm at sixteen to learn the carpenter's trade on Cape Breton Island. Working his way west as a journeyman carpenter, he had reached San Francisco in 1866 by way of Nicaragua with two brothers, Howard and Thomas.[9]

From Jack Fraser, Floyd learned of a job at the borax mines in Lake County, a hundred miles north of San Francisco. The work was hard and the wages low – somewhat over a dollar a day – but the prospect was alluring. Borax Lake, they found, lay just east of Clear Lake. The trip by steamboat up the Napa River to Napa City, then by stage to Lower Lake at the southern end of Clear Lake, took about thirty hours. It was a strenuous trip, but Dick and Jack, delighted with the country, were enthusiastic. So began a lasting love affair with Clear Lake.

But this job did not last. Soon Floyd returned to the sea, sailing first up and down the Pacific Coast to Victoria, then out to the Sandwich Islands. On one of these voyages, probably as captain of the *Idaho,* he met and fell in love with a young passenger, Cora Lyons, who was traveling with her father, Henry Augustus Lyons, a former chief justice of the Supreme Court of California. Henry Lyons had married a famous belle, Eliza Pirrie, at Oakley Plantation in St. Francisville, Louisiana.[10] The small, dark-haired, dark-eyed Cora was captivated by Dick's enthusiasm, his boundless energy, and spirit that matched her own.[11]

Cora Lyons Floyd. (Reproduced by kind permission of the California Historical Society.)

In September 1871 Richard Floyd and Cora Lyons were married at Trinity Church in San Francisco.[12] They moved into the "palatial" house at 415 First Street on fashionable Rincon Hill overlooking the ship-filled harbor (Cora's father had won the house in a poker game from the famous attorney Hall McAllister).[13] Lyons gave this house along with $20,000 as a wedding present to his daughter. After this Floyd made one more voyage as captain of the *John L. Stephens,* an ancient side-wheeler thought by many to be doomed.[14] When, after weathering a wild storm off Astoria, Floyd returned to his bride in San Francisco, she urged him to give up the sea. Soon after this, Cora's father died, leaving her a fortune in real estate, and her urging turned to insistence. She was pregnant and wanted her husband near. The baby proved to be a girl. Named for her grandfather, she was called Harry or Hal, using the initials of her given name, Harry Augustus Lyons Floyd.[15] Cora, happy with her baby and her gay, absorbing life in San Francisco, wanted nothing more. But was Dick Floyd really ready to settle down to a life of pleasure, away from the sea, living off his wife's fortune and managing Henry Lyons's estate? He too must have wondered.

As we have seen, Lick's First Deed of Trust was dated July 16, 1874, and Floyd's first meeting with the old man took place in that same year. In the months that followed, Lick grew increasingly impatient with his first board of trustees. He was enraged to find that, according to his First Deed, he could not change the trustees or the terms of the deed, nor could he revoke the Trust. He confessed he had made the deed "hastily . . . under the effects of mental depression, caused by failing health and apprehension of a speedy departure from this life." On March 24, 1875, he wrote to the trustees to tell them angrily of these "mistakes that would postpone the execution of the great works he had contemplated until after his death," a result he never intended.[16]

One day soon after this, Lick turned to Tom Fraser and exclaimed, "I like that young man Floyd," in his usual abrupt, straightforward, and conclusive way. He then said, "I would like to make him a Trustee and have him take charge of the observatory. Go and bring Selby here, as I want him to resign and put this man Floyd in his place." To which Fraser added, "Mr. Lick always after – spoke very highly of Captain Floyd and considered him just the man that suited him – to manage the building of the telescope." [17]

But Thomas Selby, after first agreeing to Lick's demand, refused to resign. The other trustees supported him and threw the resignation into the fire. Lick was furious.

The old man had weathered many fights over the control of his lands. Once he had nearly been shot in the courtroom during a trial stemming from a quarrel over titles.[18] Now he consulted his attorney, Theodore H. Hittell, and then offered the brilliant lawyer John B. Felton $100,000 to represent him in a case against the trustees. Felton, knowing the power of the San Francisco press, decided to present his case there. He suggested that the public, eager to see Lick's Trust executed and the observatory completed, had long been wondering why the trustees were so slow in carrying out Lick's desires. Soon, to Felton's delight and Lick's amazement, the trustees capitulated. They all resigned and asked to be absolved from the Trust. Thus, far more easily than he expected, Lick had his way. Still he hated to pay the $100,000 fee. Felton reminded him of the story of the dentist who treated a farmer with a bad toothache. He seated his patient on a rickety chair, and with a set of rusty instruments he extracted the tooth easily, painlessly. The charge was $5. The farmer objected. "Five dollars!" he cried. "Why, when Jones down at the village pulled my last tooth, it took three hours, during which he broke his chair, broke my jaw, broke his tools and he only charged a dollar. You ought to cut your bill." [19]

When Felton finished his story Lick gazed at him for some minutes and said nothing. Finally he said grimly, "I don't think it is a parallel case, but I guess we better pay the bill." Felton got his $100,000 plus $36,000 for his legal costs.

Lick was now free to choose a new board of trustees. To Floyd's surprise and his wife's initial delight, he chose Floyd as president of this, the Second Lick Trust. The *Napa Reporter* heralded Lick's choice in an extravagant article Cora cut out and pasted in her scrapbook. The *Reporter* called Floyd "one of the most thorough mathematicians and practical scientific men in California." Mr. Lick, it said, "could not have placed at the head of his trustees a man more thoroughly capable of carrying out his bequests in the interests of Science than Richard S. Floyd."[20]

A former classmate at Annapolis, hearing the news, congratulated his friend: "I do not forget your bright promise as a boy, and I am happy to see that in your manhood you have more than realized that promise – that you should 'turn up' in a telescope would indeed surprise me, except that the certain, or uncertain amount of mathematics to be applied is illimitable – whereat and in you are no doubt happy." He recalled Floyd's droll sketches and said he would have looked more for another Doré or Nast.[21]

Some months after the formation of the Second Trust, Floyd went down to see Henry E. Mathews in his office on California Street. Mathews had been the bookkeeper for Isaac Friedlander, the Wheat King, Henry Lyons's close friend and the executor of his estate. Floyd asked now if he knew a young man who might qualify as secretary for an important board of trustees. Mathews, who had known Floyd in his seafaring days, realized that this was an indirect way of asking him to serve on the board. He offered his services, and the appointment was seconded by Faxon D. Atherton, treasurer of the Second Board, who had known Lick in Chile.

In contrast to the towering 300-pound Friedlander, who strode along the docks in his Prince Albert coat and stovepipe hat, Mathews was a gentle, dapper little man, who, through his boss, learned to manipulate money in wide-ranging, complex wheat deals with the British. After leaving Friedlander in 1875, Mathews welcomed the chance to become secretary of the Lick Trust, even though the Lick office turned out to be a dingy closet in the Lick House, "afterwards appropriately used by the porters and bootblacks."[22]

A native of Illinois, Mathews had arrived in California in 1855. A member of the first class of the first high school in San

Francisco, he had worked for his father before joining Friedlander. For the Lick Trust, this choice proved to be an excellent one. A methodical, conscientious man, Mathews loved to keep records and write letters; he was also a skilled photographer and a Greek scholar.[23] As the observatory grew out of the wilderness, he became a vital part of the project, recording every phase of its history with pen and photographic plate.[24]

The first question was that of the observatory site. Early in 1875 Fraser climbed Mount St. Helena and found two acres of level ground but no water within 1,000 feet of the top. Lick, concerned about his enterprise and eager to see the place for himself, asked to be taken there. He journeyed there on a mattress in a horse-drawn wagon. He visited Calistoga Hot Springs, a health spa, and "camped out" about a third of the way from the top. Unfortunately, on the way up the rutted road, the wagon toppled and Lick and his mattress landed unceremoniously in the dust. There he lay, "cursing all Napa County." Afterward he said he did not like the St. Helena site very well, yet he supposed it was the best they could do. Floyd suggested other sites – Mount Diablo and Loma Prieta and Mount Bache in the Santa Cruz Mountains. Lick, however, was not enthusiastic about any of them.[25]

Fraser then had a brilliant idea. Not far from Lick's homestead in Santa Clara County there was a mountain in the Diablo range, about twenty-five miles east of San Jose, that reached a good height for an observatory – 4,400 feet. It was called Mount Hamilton and Lick could see it from his mill at Alviso. Nothing could have pleased the old man more.[26]

In the early days, this mountain, once called the watchtower of the Indian, had been named La Sierra de Ysabel. Here the notorious bandit Joaquin Murieta, after ravaging the villages in the valley, found refuge in a mountain gorge. On August 28, 1861, William H. Brewer, a geologist in charge of the field corps of the State Geological Survey, made the first recorded ascent, accompanied by Charles T. Hoffman, topographer of the State Survey, and Laurentine Hamilton, a Presbyterian minister in San Jose.

The Reverend Laurentine Hamilton, for whom Mount Hamilton was named. (Reproduced by kind permission of the Mary Lea Shane Archives of the Lick Observatory.)

Starting out on mules, they found many bear tracks but no trails through the heavy underbrush. At last they picketed their mules and struggled up the last three miles on foot. Hamilton pushed ahead of his companions, who were heavily laden with equipment. When he reached the summit, he swung his hat in the air and cried out, "First on the top, for this is the highest point." Brewer decided then and there to call the peak after his "noble and true" friend. Thus the minister gained immortality.[27]

At 10:00 A.M. on August 31, 1875, fourteen years after Brewer's ascent, Tom Fraser and Bernard D. Murphy, mayor of San Jose and a Lick trustee, arrived on horseback to explore the site. Despite the wild nature of the region – the chaparral and mesquite, the thickets of scrub oak covering the slopes, the absence of a trail – Fraser was taken with the site. He found only two serious drawbacks: First, there was no water near the summit; second, the sharp peak would have to be leveled by blasting some thirty feet of graywacke off the top. But Fraser saw that, despite the rough terrain, a road could be built, and so, as he was to write years later, "I felt as confident that the Lick Observatory would be erected as I do now after work has begun in earnest." On September 1 he reported to Lick in San Francisco. The millionaire, elated, said right away that if a road could be built and the mountaintop proved suitable, Mount Hamilton would become the site of the largest telescope in the world.[28]

News of Lick's choice spread rapidly. *Harper's Weekly* quoted the *San Jose Mercury:* "Mt. Hamilton is 4,448 feet high. The summit is higher than any land within fifty miles. . . . The magnificent valley of San Jose, the snowy ridge of the Sierra Nevada, and a boundless view of mountain scenery are in the scope of vision, and the elevation is so high as to be above the fogs of summer, and is not so high as to be much disturbed by the storms of winter. . . . The road will open up a rich territory of agricultural lands, besides furnishing one of the most delightful and romantic drives in the United States."[29]

There were objections, of course. Newcomb had "grave doubts." He feared currents of warm air rising up the mountainsides and enveloping the top might be fatal to astronomical "seeing." No large mountain observatory had ever been built in the United States, and so he had good reason to doubt. Long after, Edward Holden wrote in retrospect, "The possibility that a complete astronomical establishment might one day be planted on the summit seemed more like a fairy tale than sober fact."[30]

III. OPERATIONS UNDER THE PROVISIONS OF MR. LICK'S SECOND DEED OF TRUST.

Mr. LICK'S second deed of trust was dated September 21, 1875, and appointed the following Trustees:

RICHARD S. FLOYD, esqr., JOHN H. LICK, esqr.,
FAXON D. ATHERTON, sr., esqr., JOHN NIGHTINGALE, esqr.,
BERNARD D. MURPHY, esqr.

This Board of Trustees was succeeded by a third Board (the present one), under appointments made by JAMES LICK September 2, 1876; ratified by action of the Trustees in session of the Board November 29, 1876, and confirmed by action of the Courts during the settlement of compromises with the heirs of Mr. LICK after his decease. The present board is composed as follows:

RICHARD S. FLOYD, esqr., President;
WILLIAM SHERMAN, esqr., Vice-President (died September 12, 1884);
EDWIN B. MASTICK, esqr., Treasurer;
CHARLES M. PLUM, esqr.,
GEORGE SCHOENWALD, esqr.

The section of Mr. LICK'S last deed under which the Board is acting is given below.

EXTRACT FROM MR. LICK'S SECOND DEED OF TRUST.

"THIRD—To expend the sum of seven hundred thousand dollars ($700,000) for the purpose of purchasing land, and constructing and putting up on such land as shall be designated by the party of the first part, a powerful telescope, superior to and more powerful than any telescope yet made, with all the machinery appertaining thereto and appropriately connected therewith, or that is necessary and convenient to the most powerful telescope now in use, or suited to one more powerful than any yet constructed; and also a suitable observatory connected therewith. The parties of the second part hereto, and their successors, shall, as soon as said telescope and observatory are constructed, convey the land whereupon the same may be situated, and the telescope and the observatory, and all the machinery and apparatus connected therewith, to the corporation known as the 'Regents of the University of California;' and if, after the construction of said telescope and observatory, there shall remain of said seven hundred thousand dollars in gold coin any surplus, the said parties of the second part shall turn over such surplus to said corporation, to be invested by it in bonds of the United States, or of the City and County of San Francisco, or other good and safe interest-bearing bonds, and the income thereof shall be devoted to the maintenance of said telescope and the observatory connected therewith, and shall be made useful in promoting science; and the said telescope and observatory are to be known as 'the Lick Astronomical Department of the University of California.'"

Operations under the Provisions of Mr. Lick's Second Deed of Trust, 1887. (From *Publications of the Lick Observatory,* vol. 1).

In his Second Deed of Trust, dated September 21, 1875, and said to have been drawn up by Felton, Lick dictated that, when the observatory was finished, it was to be turned over as the Lick Astronomical Department of the University of California to the board of regents, of which, incidentally, Felton was a member.[31]

Apparently Lick, persuaded by Felton, was sure the State of California would support the telescope handsomely as a matter of pride, if for no other reason. Floyd had misgivings about the affiliation of the observatory with the university regents, a dependence he considered very uncertain. He was afraid the same mistake might have been made that had prevented the usefulness of most American observatories. He told Newcomb, "It is impossible to tell to what special line of work the Observatory will be devoted – never mind for what special line we might intend it. Should the University continue under as conservative a management as at present it might do well with such a costly Astronomical Department. But every time there is a session of the California Legislature the politicians make bold efforts to get its control and if they ever succeed I fear we shall have a new style of scientists in political astronomers, and that a great observatory might be thrown away."[32]

Floyd, in one of his first acts as president of the Second Trust, wrote to the regents to tell them the terms of the deed by which $700,000 was to be given for the purchase of land, on which a telescope superior to, and more powerful than, any yet built would be mounted, with a suitable observatory connected to it. "We shall be pleased to hear that you accept the Donation; and also in what manner and to what extent you authorize to assist in carrying out the views of the beneficent."[33]

The regents answered, "In response to this act of spontaneous, and almost unparalleled munificence whereby the interests of science, through one of its most exalted departments, may be able to attain upon the shores of the Pacific and in one of the youngest states of the Republic, a higher advancement than has ever yet been reached by the oldest and most enlightened nations of the globe, the Regents of the University have delegated to us the duty of announcing their acceptance of this splendid gift and their appreciation of the noble purpose of its author."[34] They appointed three members of their board to consider the problems and agreed to cooperate in gaining title to the Mount Hamilton lands designated by Lick.

Lick was, of course, delighted by this recognition of his "munificence." But the change from his plan, outlined in the First Deed, and his designation of the university as recipient caused widespread reverberations. Benjamin Alvord protested the new plan, "instead of the simple efficient plan of the old Trustees . . . to appoint an astronomer at once and build it up afterward in a *plan* with forethought and comprehension of the facts."[35]

But Alvord's anger did not compare to that of George Davidson, who on his return from his world tour, learned that the $1 million plus he had proposed for the observatory had been reduced to a paltry $700,000. The 10,000-foot site he had advocated in the High Sierras had been abandoned in favor of Mount Hamilton, a site "vastly inferior . . . a gratification of personal vanity." Lick had flagrantly ignored his expert advice. In his fury Davidson refused to have anything further to do with Lick or his observatory. If he dreamed of becoming director of the new observatory, he may have realized that this was now highly unlikely. It was well known that Lick liked to make up his own mind, hated to be told what to do. Therefore, although Davidson had encouraged the old man's faith in a telescope as a suitable monument, he may have pushed his own views too far, thereby thwarting his personal dreams. A few years later he would console himself by building a small observatory with a 6.4-inch telescope at the corner of Laguna and Sacramento streets in Lafayette Park in San Francisco. Through the press he became one of the chief critics of those appointed to build Lick's observatory.[36]

For Dick Floyd, on the other hand, a new and different life began as he joined Tom Fraser in the building of the world's greatest telescope on a remote mountain peak. It was a formidable job. Yet, although neither man was an astronomer, and neither had any real astronomical training, they never doubted their ability to carry out Lick's desires, despite the many obstacles, human and mechanical. The job would take longer and be harder than either anticipated. At the end it would cut both their lives short. For the ingenious, skillful, hardworking Fraser it meant total devotion to the project. For the flamboyant, energetic Floyd it meant the partial abandonment of the life of "Amusement" he had looked forward to with Cora at Clear Lake.[37]

On their honeymoon Cora had fallen in love with this crystal-clear lake where fish abounded, and with the valleys and mountains where game of all kinds flourished – where even bear still roamed. Once, after a "glorious day's outing," Dick wrote enthusiastically to a friend, "We are all well! The weather is splendid. So are the geese and the ducks. We are having splendid sport. . . . Mrs. Floyd, Jack [Fraser] and I got a lot of big Mallards, wood duck and teal yesterday. We have the trick for ducks now. Bait the stands with wheat. You ought to see Mrs. F. make the feathers fly!"

Here, in 1874, about the time Floyd and Lick met, the Floyds bought nearly 300 acres on a triangular peninsula extending into the lake where it narrows – at the western end of Paradise Valley at the foot of Floyd Mountain (as it would be called). Across the lake rose the extinct volcano Konocti (popularly known as Uncle Sam). Here they planned a house they would call "Kono Tayee," said to mean Mountain Point. The house was designed in the form of a ship, and Jack and Tom Fraser would build it of redwood. The job would not be easy; with no road on the north side of the lake, all the building materials would have to be transported by barge from Lower Lake.

When the house was finished, Floyd called it the "lovelyest place in the world."[38] Visitors wrote of the beautiful situation of the house, its warm and friendly atmosphere. A frequent guest wrote, "Floyd's hospitality was unlimited; his open eyed welcome with a gracious, engaging and gracious look, his heart frank, honest and expansive, seemd gratefully to meet you." A great raconteur, fond of a good story, a lover of music, Floyd knew how to entertain his guests. He enjoyed himself hugely, "in jovial or serious company." Friends came from far and near to join in the hunting and fishing and to revel in lavish parties, complete with fish and venison and the best French wines. A relaxed host, he let his guests do as they pleased, even if, at times, this led to excesses.

To Floyd Kono Tayee seemed perfect – with its splendid water facilities, provided by two large reservoirs behind the house supplied by a mountain brook with 80,000 gallons of sparkling water a day; the pressure at the house, created by a "105 feet fall;" the fountain with a nozzle in the center that in still weather shot water seventy-five feet high. Best of all, the abundance of water enabled Floyd to irrigate the grounds – the monkey, magnolia, palm, orange, acacia, pepper, and pomegranate trees brought from the far corners of the earth.

Now, however, as he plunged into work for the Lick Trust, Floyd had to sacrifice this carefree life. Instead of making steamer models for his own and his guests' amusement, he would have to apply his mechanical ingenuity and inventive skill to the design of diverse astronomical instruments and mountings, in addition to the great telescope.

On September 4, 1875, Lick wrote to the Santa Clara Board of Supervisors to ask it to build a road up Mount Hamilton.[39] His letter reveals much of his character and the nature of his dreams: "Being more interested in the Observatory and telescope, which I have ordered to be erected on this coast, than any other of my projects, and which I intend to be in advance of any other scientific work in the world, the question of its location has been of deep interest to me. Of the many locations proposed, I have after much deliberation thought favorably of locating it on the summit of Mount Hamilton, Santa Clara County, providing this my petition to your honorable body is received with favor. The advantages its near vicinity would be to the general public and tourists from abroad, of course I have taken into consideration, but above all the benefits to be derived from it by the inhabitants of Santa Clara County and especially the City of San Jose."

He then asked the county supervisors to build a road to the summit of Mount Hamilton. He insisted that the road be first class in every particular and asked them to take action without delay. To facilitate matters he offered to accept Santa Clara County bonds in payment and to advance the money for the road.[40]

The supervisors accepted Lick's petition, and work began in February 1876. The surveyor for Santa Clara County, A. T. Hermann, was engaged along with his brother to survey the contour lines and to decide how much cutting of the rocky peak would be needed to form a plateau for the observatory buildings. In mid-March Hermann wrote to Floyd, "The road is going ahead bravely," and offered the hospitality of their camp, which at one time had been buried under four feet of snow.[41] Floyd, with Fraser, accepted the invitation and spent the night in a small cabin, from which they climbed the rough trail to the top.

By April, despite the failure of the first contractor, the road was within three miles of the top. All should have been well. But Congress still had not approved the grant of United States public lands to the University of California for the observatory. Floyd wrote anxiously to Aaron Sargent, California senator in Washington, emphasizing the need for haste.[42]

By December 1876, the "hand built" highway was finished at a cost of $73,075.81. In January 1877, the supervisors and the Lick trustees, riding in a Concord coach, visited the site. From Hall's Valley they spun along the road "in and out of ravines, catching at every turn glimpses of magnificent scenery." The road, "a marvel of engineering skill and an enduring tribute to the engineers," was twenty-six miles from San Jose, twenty-two and a half miles in length from the base of the mountain. With its 365 twists and turns, the road was so beautifully graded that at no point along it did the stage horses need to break their trot.[43] It was called "Lick Avenue."

One day, while they waited to hear from Washington, Fraser had ridden up the mountain on horseback to see how the work was getting on. To his horror he discovered two squatters in a shack, claiming ownership of the land. He wrote immediately to Floyd, asking permission to build a small house on the top where he could put his trusty men in charge. Floyd replied, "Go ahead with house on the top immediately. Take possession of the top of Mt. Hamilton and hold it."[44] Fraser rushed ahead. He set fire to the squatters' shack. In its place he built a house eight by twelve feet, with door and lock. He had a hard time getting the lumber up the mountain. In many places his men, paid $2 a day, had to carry it. He installed three guards until the squatters, now camped on the other side of the mountain, moved on.

3

European journey

On a warm day in the spring of 1876 Floyd arrived in Washington, D.C., with Cora and three-year-old Harry. As president of the Lick Trust he was on his way to Europe to follow up on Simon Newcomb's trip the year before and to consult with leading astronomers and optical experts on the problems of lenses, mirrors, mountings, and domes for the Lick Observatory. He stopped first at the Naval Observatory to see the redoubtable Newcomb. The observatory stood on a hill above the Potomac River in northwestern Washington, an unhealthy region infested with malaria-bearing mosquitoes and reached by muddy roads. In an office there he found Newcomb, whose powerful head was crowned by a shock of dark hair, his face decorated with a drooping mustache and flowing sideburns, large ears, and dark blue eyes. Floyd was duly impressed. As he came to know Newcomb better, he would be even more impressed by the astronomer's breadth of knowledge and a versatility that ranged from astronomy to economics, from philosophy to linguistics, from finance to fiction.

That afternoon Newcomb showed Floyd around the observatory, into the revolving dome where the 26-inch refracting telescope, with its huge 35-foot tube, soared. The lenses of this, the most powerful telescope yet built, were so exquisitely figured by Alvan Clark that "all the light of a star gathered by the great surface is packed at the distant focus (in the eyepiece) into a circle very much smaller than that made by the dot on this i."[1] As

The U.S. Naval Observatory, about 1876, showing the dome of the great equatorial. (Reproduced by kind permission of Jan K. Herman of the Naval Medical Command.)

Newcomb talked of the telescope and his observations with it, Floyd saw vividly what an even greater instrument might reveal.

That evening, as they sat in front of a fire and Newcomb told of his beginnings, Floyd listened closely. Like the Frasers, Newcomb had come to the United States from Nova Scotia. The precocious son of an itinerant country school teacher of New England descent, he was, at sixteen, apprenticed to a notorious doctor near St. John who reportedly performed miraculous cures. This Doctor Foshay, at first meeting charming and outgoing, proved to be the exact opposite – "silent, cold, impassive," a complete humbug. After two hard years in which he learned next to nothing, young Newcomb ran away. At Calais, Maine, he boarded a ship bound for Salem, Massachusetts. There he joined his father, who since his wife's death the year before, had been

Simon Newcomb at age forty-four. (Engraving by J. J. Cade, 1879, in the *Eclectic Magazine,* vol. 31, April 1880. Reproduced by kind permission of Brenda Corbin, librarian of the U.S. Naval Observatory.)

living in Maryland. There, too, Newcomb began to devour everything he could find on mathematics, including Newton's *Principia* and Laplace's *Mécanique Celeste.* In 1857 this self-teaching led to jobs as computer with the Nautical Almanac in

Cambridge, Massachusetts, and then with the Naval Observatory in Washington.[2]

As Newcomb talked, Floyd followed eagerly his account of his research into the complicated problems of the moon's motions – work that took a special mathematical genius. Writing of the person who is moved to the exploration of nature by a dominating passion, Newcomb would state, "If he is destined to advance the science by works of real genius, he must, like the poet, be born, not made. The born astronomer, when placed in command of a telescope, goes about using it as naturally and effectively as the babe avails itself of its mother's breast. He sees intuitively what less gifted men have to learn by long study and tedious experiment. He is moved to celestial knowledge by a passion which dominates his nature. He can no more avoid doing astronomical work, whether in the line of observations or research than a poet can chain his Pegasus to Earth."[3]

If Newcomb was an idealist in his chosen profession, he could, as Floyd would learn, also be difficult. Forthright and honest – at times painfully so – he loved truth as he saw it and hated sham. He said exactly what he thought and always did what he considered right.

As they talked, Floyd must have sensed that the astronomer was not only puzzled by, but questioned strongly, his selection by Lick for the presidency of the Lick Trust. The ability to navigate a ship and use a sextant did not mean he could guide the building of a great telescope. Nevertheless, Newcomb told Floyd of his meeting with D. O. Mills of the First Lick Trust and summarized the ideas on the lens, mounting, and observatory site brought out in their conversation.[4] These had covered the kind of glass to be ordered, the time needed to cast the glass and figure the lenses, and the construction of the mounting for the 36-inch telescope. Afterward Mills, a University of California regent, had talked with President Charles Eliot of Harvard and the observatory staff there. All had expressed interest in "our great project."[5]

The chief question was this: Should Lick's great telescope be a refractor, like that in Washington, focusing the light with a lens, or a reflector, which used a mirror for this purpose? This question would plague the trustees for months to come. Although the advantages of a large reflector were great, the mechanical problems involved in supporting a large mirror were formidable. As a result, most astronomical work to date had been done by refractors.

Newcomb told Floyd of his visits to observatories and to optical and instrument shops in England, France, and Germany in 1875; he then showed him his report with the information he had obtained that would help the trustees "shape their course."[6] They discussed all the questions that still had to be answered. Newcomb then gave him letters of introduction to leading European astronomers and optical firms and wished him well on a trip that would last longer than a year.

Just before leaving Washington Floyd learned that the Mount Hamilton bill had finally passed both houses of Congress "all right." The land granted by the government for the observatory embraced a circle a mile around below the summit.[7]

In New York Floyd called on Henry Draper, formerly a doctor at Bellevue Hospital, who had been taking "quite an interest" in the start of the Lick Observatory.[8] After building his own little observatory at Hastings-on-Hudson, Draper had become increasingly fascinated by spectroscopy and photography and their application to analysis of the physical nature of the sun and stars.

This field, as Joseph Henry had explained to Lick, contrasted strongly with the "old astronomy," concerned with the positions and motions of moon, planets, and stars, in which Newcomb was working.[9] Instead of depending on mathematical calculation, the new astronomy, or "astro physics," as Henry called it, relied on the study of that small band of colored light known as a spectrum. As the astronomer turned his spectroscope on sun and stars, he saw that spectrum crossed by thousands of lines, each "broadcasting" on a different wavelength, each telling the story of the physical composition of an individual star. When he compared the stellar lines with lines from various elements in his laboratory, he found he could identify iron, cobalt, and magnesium. "It is as though a star throws the whole secret of its being into its spectrum, and we have only to read it aright in order to solve the most abstruse problems of the physical universe."[10]

Draper was now absorbed in this new field. With his 28-inch reflecting telescope he would be the first to obtain a satisfactory spectrum of the brilliant star Sirius. In 1867 he had married the daughter of the wealthy Courtland Palmer, the beautiful auburn-haired Anna Palmer, who became his devoted assistant and in time a benefactor to astronomy.

With Floyd, Draper talked of his 12-inch refractor, made by Alvan Clark, and its advantages. But he talked, too, of the promise of his reflecting telescope in the field of "astrophotography." Amazed by the incredible sensitivity of the photographic plate, he believed astronomers would soon be able to photograph what they could not see. Yet, while he realized the advantages of the reflector over a refractor, especially for photography, he was, at this time, uncertain which to recommend for the Lick telescope.

One day Newcomb joined them in New York. At the time he thought it might be impossible to cast the disks for a 34-inch refractor, one larger than any then in existence. Therefore he favored a large silvered-glass reflector.[11]

In mid-May, Captain Dick Floyd with Cora and little Harry sailed out of New York harbor. He was happy to be at sea again, happy to be leaving behind the turmoil of the 1876 presidential election, a bitter contest between Hayes and Tilden. More than ten years had passed since he had left France after the war. Sixteen years had passed since Cora had lived abroad with her father and sister Lucy. It was a different Europe they now entered. Instead of living the life of a searaider, Dick would spend his time visiting observatories – from London to Edinburgh, from Paris to Nice and Vienna. Everywhere this tall, dark American, so different from many of the older astronomers he met, asked questions, trying to solve the problems he faced. Whenever he could find the time he sent back reports on his findings.

Long afterward the historian Howard Sonenfeld, writing of James Lick and the history of California, described Floyd's activities on behalf of the Lick Trust: "Richard Floyd worked like a Trojan, collecting information on how to construct the greatest telescope in the world. Working at an unbelievable pace, he travelled all over Europe, running down every lead, contacting any man who might even remotely assist in the enormous project."

As Floyd traveled, meeting scientists and engineers of many kinds, he learned much, not only of their personalities but also of the growth of astronomy in this dramatic period when, with advances in photography and the development of spectroscopy, it was undergoing a revolution. Even one not trained as an astronomer could see the immense possibilities for the exploration of the universe in the years ahead. With characteristic energy and enthusiasm, Floyd was to do all he could to learn of these possibilities and apply them to the technical problems of building a great telescope in fulfillment of James Lick's Trust.

A few months earlier a leading American physicist, Henry A. Rowland of Johns Hopkins University, had visited various European observatories. In general he found them "far inferior to many in America." Yet he considered their work "superior in quantity and *perhaps* in quality to most of it in our country." [12] Now Floyd had to make his own observations and draw his own conclusions. Everywhere he went he was flooded with invitations from those who wanted to show him their telescopes and give him the best information they could.

In Dublin he saw Howard Grubb, the son of Thomas Grubb. Howard, a self-taught mechanic, had designed machinery for printing bank notes for the Bank of Ireland, before turning to the manufacture of machine tools and telescopes. He had studied civil engineering at Trinity College in Dublin. In 1865 he had left college to help his father with the great 48-inch reflector for the observatory in Melbourne, Australia. He was now carrying on his father's business at Rathmines, outside Dublin. Floyd visited the factory there and had a good chance to see the work in progress, as Grubb was working on the 27½-inch refractor for the Vienna Observatory.[13]

The flint disk had just arrived from Charles Feil in Paris. It was "perfectly white and beautiful." But, as Floyd would learn from hard experience, this was the easiest part of the objective to cast.

The crown glass for the convex lens needed to make the object glass achromatic, bringing all the colors to a common focus, proved much harder to make.

In Dublin Newcomb talked with Grubb about the construction of a great reflector. They discussed the contract terms for such a telescope, and Grubb even offered to make an experimental mirror. But Lick declined his offer.

Newcomb called Grubb a genius who meant business, although he had not yet produced a work that would justify unlimited confidence in his abilities. Some astronomers disagreed. They claimed he had the instincts of a mechanic, making pretense to all sorts of secrets he could not divulge.[14] Floyd found him intelligent and ambitious, "an excellent optician and mechanic" who would offer many vital suggestions for the Lick telescope.[15] He was a warmhearted, delightful person and the two men, who were about the same age, had several happy meetings, with "pleasant little chats" to which Grubb would refer afterward, noting Floyd's "many marks of kindness." Both Dick and Cora Floyd, in turn, appreciated Grubb's friendship and that of his wife, the daughter of a Louisiana doctor. Floyd appreciated also the letters of introduction to astronomers in England and Scotland, and the concern Grubb showed as they prepared to leave for England. It was blowing "Great Guns," and the barometer was plummeting. Grubb sent an anxious note. If they could delay their departure twenty-four hours, he thought Mrs. Floyd would be better pleased.[16] The Floyds waited.

Grubb, of course, hoped that out of his meeting with Floyd would come the chance to build the great telescope. He wrote afterward, "I am quite willing to do it for the glory of it; for if this instrument is put in my hands I intend that it shall be the great work of my life and on it my reputation will stand or fall."

Before leaving Ireland, Floyd had gone to see Lord Rosse's huge reflecting telescope, the "leviathan," swinging between the gates of the park at Birr Castle near Parsonstown. He compared its tube, fifty-six feet long, six feet in diameter, to the funnel of a large steamship. With this huge, if cumbersome, instrument, Rosse managed to observe thousands of galaxies, resolved some

Lord Rosse's reflecting telescope, 6-foot aperture. Photograph taken at Birr Castle, Ireland, "inscribed to the Lick Trustees from Ralph Copeland," dated June 29, 1876. (Reproduced by kind permission of the Mary Lea Shane Archives of the Lick Observatory.)

of them into individual stars, and suggested the spiral forms of others. Soon after his arrival in England, Floyd met Ralph Copeland, who for many years had operated this telescope and favored "enthusiastically" a speculum (or metal alloy) reflector for the Lick telescope.[17]

Outside London, near Dulwich, at 90 Upper Tulse Hill, Floyd found the noted English astronomer whom Grubb called "the Prince of Physical Astronomers."[18] William Huggins, like Henry Draper, was working in the forefront of the new astronomy to discover the nature of stars and nebulae. In the observatory behind his house he had mounted the 15-inch Grubb refractor and an 18-inch Cassegrain speculum reflector and combined them with a spectroscopic laboratory. As he talked ardently of the work he had done and his hopes for the future, Floyd shared his excitement. "The time," Huggins was to write, "was indeed one of strained expectation and of scientific exaltation for the astronomer almost without parallel; for nearly every observation revealed a new fact, and almost every night's work was red-lettered by some discovery." [19]

Just at this time, using a dry gelatin plate with his reflecting telescope, Huggins succeeded in photographing the spectrum of the blue star Vega. In it he discerned seven strong lines – a remarkable achievement, as the light received at the earth from such a first-magnitude star is only about one forty billionth that received from the sun.

Like Henry Draper, Huggins was helped by his charming wife, Margaret Lindsay Murray, who was from Dublin. When they had married the previous year, he had been fifty-one and she twenty-six. She had, she said, loved the stars since childhood. In her teens she had charted the sunspots with a little homemade telescope and had observed the lines in the solar spectrum with a small spectroscope she had devised. She had even experimented in the new field of photography. Like her husband, a man of broad interests who played the violin and owned a Stradivarius, Margaret Huggins was talented musically and artistically. They welcomed Floyd warmly. Long afterward Huggins remembered "Captain Floyd's visit for consultation on the observatory he was about to found." [20] He predicted then that in the future the telescope of discovery would be a large reflector. But as a result of the apparent inability of most astronomers to use a reflector successfully and the "failure of mechanics to preserve permanent stability of collimation," he urged the Captain to choose a large refractor as the Lick telescope.

Meanwhile, west of London, at Maidenhead, Floyd visited William Lassell, brewer turned astronomer, who had built a 4-foot speculum reflector, with an ingenious lever support system he had invented. The old man was optimistic about the future of reflectors. Yet, like Huggins, he recognized the difficulty of making suitable mountings and considered a 48-inch reflector the largest instrument then feasible. Once he even said pessimistically that he feared the telescope as an instrument for scrutinizing the heavenly bodies had reached its culminating point![21]

In Birmingham Floyd then went to see the makers of the optical disks for the Naval Observatory. Newcomb considered this old and wealthy optical firm, Chance and Company, with Feil of Paris, one of the two companies that promised success in the production of such disks. He considered Chance far more responsible than Feil, who was comparatively "little more than an adventurer," with more enthusiasm than capital. Yet Newcomb felt that Feil had recently had better success producing large disks.[22]

Back in London Floyd talked with George Stokes, the physicist whom Grubb called "our first science man in England." At the South Kensington Museum he called on Huggins's protagonist, Norman Lockyer, the astronomer whom Joseph Henry had proposed as director of the Lick Observatory. A brilliant man, overflowing with new, often radical ideas, Lockyer was the founder and editor of the journal *Nature*. In 1868, with a spectroscope attached to his telescope, he had made the dramatic observation of the huge flamelike prominences rising from the sun's edge – the first time they had been seen outside of a solar eclipse. Almost simultaneously the French astronomer Jules Janssen made similar observations.

Here, at the museum, Floyd called on Newcomb's assistant, Edward S. Holden, who had been sent by the United States government to examine and report on the Loan Collection of Scientific Instruments there. Newcomb had written to Holden, "President Richard S. Floyd is probably bobbing about London and elsewhere. Keep a look out for him."[23] Their meeting was brief, and at the time Floyd thought little of it. Later he would write, "I then barely made his acquaintance and conversed with him generally upon astronomical matters as I did with every man I met who took an interest in the subject."[24] After this, Floyd went on to Paris.

An article in *Nature* reported his arrival: "Mr. Floyd, the president of the trustees of the Lick donation for the construction and fitting up of the San Francisco Observatory, arrived in Paris at the end of June. His first visit was to M. Leverrier, who

Richard S. Floyd in Paris, France, 1876, at age thirty-three. (Reproduced by kind permission of the Mary Lea Shane Archives of the Lick Observatory.)

gave him every assistance in his power to enable him to fulfill the object of his mission. Mr. Floyd is at liberty to use the observatory grounds for any experiments in connection with his large refractor, which it is intended to construct. M. Leverrier concurred with him in not attempting to construct a lens of more than one metre in diameter. The money at the disposal of Mr. Floyd is 200,000 pounds."[25]

Meanwhile Newcomb, concerned over Floyd's progress, wrote anxiously to Holden, "I hope you have heard from Grubb and others what President Floyd is doing in Europe. I see a mention of his visit to Paris in *Nature* of July 13."[26] The notice read, "Mr. Floyd, the President of the Board of Trustees of the Lick donation, has come to an arrangement with M. Leverrier for the better execution of the contemplated instrument for the Paris and San Francisco Observatories. The masses of glass are to be made in Paris."[27] Evidently doubtful of Floyd's ability to handle the presidency of the Lick Trust, and fearful of his impetuosity, Newcomb added, "Lick had better recall him before he does mischief."[28]

Floyd had indeed been in Paris in June. Anxious to learn how much crown and flint disks for a 30- to 40-inch refractor might cost, he had visited the optical firm of Feil and talked with its director, Charles Feil, at his foundry at 36, Rue Le Brun, between Boulevard St. Marcel and the Avenue des Gobelins in the Latin Quarter, on the south side of Paris near the Seine. But he had not made any final agreement on the "masses of glass."

At the Paris Observatory Floyd had talked with the sixty-five-year-old director, Urbain Jean Joseph Leverrier. A native of Normandy, Leverrier, a thin, pale-faced, clean-shaven man, with gray hair and an ascetic mien, had manufactured tobacco products before turning to chemistry and mathematics. He had gained astronomical fame by his coprediction of the existence of the planet Neptune. Floyd got along well with the crusty astronomer, of whom Cornu had written, "I don't know whether Mr. Leverrier is the most detestable man in Paris, but I know he is the most detested man." Leverrier talked with Floyd of the 29-inch reflector, recently received from Adolphe Martin, the optician

Alvan Clark had proposed bringing from France to California to finish the Lick telescope at the observatory site. Disgusted with Martin's work, Leverrier had ordered the glass for a new mirror from St. Gobain. But it would take three months to make, two years to figure, three years before he could know if it would be successful. He spoke of the possibilities of a speculum metal reflector and called Lord Rosse's telescope a failure, as 4 feet only of the 6-foot diameter could ever be used. He concluded, "Whether a reflector of any kind of six feet diameter can be made a success, no one knows." To qualify as the largest telescope in the world the Lick reflector had to be larger than six feet in order to surpass the largest existing telescope of this type.

While he was in Paris in 1875 Newcomb had visited the firm of Martin and F. W. Eichens, which had recently mounted the 13-inch reflector in the Paris Observatory.[29] He considered Eichens's work the finest he had seen in Europe, combining the good qualities of German and English machinists "with a kind of native genius which is about equally likely to turn up in California, New England or Paris." But, when Martin and Eichens were asked for a price for a reflector for the Lick telescope, their answer was obviously suggested "by the supposed liberality of an eccentric California millionaire." "It was quite clear that the prospect of lightening the burden of a successful gatherer of California gold, anxious to get rid of a large surplus accumulation, was higher in their minds than the scientific glory they might acquire by constructing the largest telescope ever made." Floyd did not visit them.[30]

Still dissatisfied with his findings, Floyd continued his travels. In early September he attended a meeting in Glasgow of the British Association for the Advancement of Science. There he talked to Howard Grubb and gave him a rough drawing of Mount Hamilton. He also listened to an address by the president of the mathematics and physics section, Sir William Thomson, who had just returned from America. "I came home," Thomson said, "vividly impressed with much that I had seen both in the Great Exhibition in Philadelphia and out of it, showing the truest scientific spirit and devotion, the originality, the inventiveness, the patient persevering thoroughness of work, the appreciativeness, and the generous open-mindness and sympathy, from which the great things of science come."[31]

Thomson told of hearing Elisha Gray's "splendidly worked out Electric Telephone," and in the Canadian Department he listened to "To be or not to be . . ." through an electric telegraph wire, the invention of young Alexander Graham Bell, of Edinburgh, Montreal, and Boston.

But Thomson was most fascinated by a conversation he had had with Simon Newcomb in Joseph Henry's drawing room at the Smithsonian Institution. The subject was precession and nutation. "Disturbed by Newcomb's suspicions of the earth's irregularities as a time keeper, I could think of nothing but precession and nutation, and tides and monsoons, and meltings of the polar ice." The speech he subsequently gave in Glasgow was entitled "Evidence Regarding the Physical Conditions of the Earth."[32]

Other scientists Floyd saw at the Glasgow meeting were the physicists James Clerk Maxwell and George Stokes, secretary of the Royal Society, the geologist Archibald Geikie, and the French astronomer Jules Janssen, who had recently shown some of his beautiful solar photographs at a meeting of the Académie Française in Paris, to which Floyd had been invited.

After this, in Edinburgh, Floyd called on the director of the Royal Observatory, Henry Piazzi-Smyth, a "curious crotchetty individual," for whom Grubb had made a 2-foot silver-on-glass reflector that was in miserable condition.[33] They talked of his pioneer high-altitude observations on the towering peak of Tenerife, but Floyd found him more interested in the Great Pyramid of Egypt than in contemporary astronomy. It was too bad that Lick, with his dream of mounting a pyramid in San Francisco's center, was not there to talk on a subject so dear to his heart.

In Edinburgh Floyd also met David Gill, former director of Lord Lindsay's observatory at Aberdeen, who was fascinated by the question of the "San Francisco monster." Gill, hearing of Floyd's hectic travels, had commented that the Captain was "taking *everybody's* opinion on the respective advantages of

Refractor and Reflector," in which case, he remarked, "I fear, poor man, he may be driven to end his days in an insane asylum." [34] After a long talk, Gill made a lengthy synopsis of their conversation. He discussed the merits and demerits of the different forms of telescope and concluded that the refractor would probably remain, for a long time, the favorite instrument of the professional astronomer. He felt, too, that it would be the best suited to Mount Hamilton, where violent winds often blow. To meet the donor's wishes fully, he proposed the largest possible refractor for use at the summit, with a seven-foot reflector to be located in a sheltered place.[35]

As Gill observed, "Thus it is, and very properly has been the object of regular observatories to possess a good Refractor, and it has been left to men of special mechanical genius, Herschel, Rosse, Lassell, Draper, Nasmyth etc, to make and employ large Reflectors." [36]

In Edinburgh Floyd took time out from his hectic schedule for an operation to alleviate a bothersome old thigh injury. Forced to stay in bed, he was able, as he recovered, to write some long-overdue letters. One that had nothing to do with the Lick Trust was to his sister Rosalie, who was in despair over her financial difficulties, and his sister Mary, who was in equally dire straights. "Grasping at any straw," they had urged him to visit the maternal ancestral home, where their great-grandfather Sir John Boog had lived, and also the Hazzard estate in Bristol. There, so rumor had it, the estate in 1862 had been worth $16 million. Somehow they hoped to grab a bit of that fortune. Aunt Mary Hazzard Hamilton of Savannah thought a peerage might be gained. To help Dick in his search, his sister sent a copy of the family tree and a picture of Dunrobin Castle, seat of the duke of Sutherland. If he made any search or if he considered the whole thing a wild goose chase is not known. In any case, his sisters never "struck it rich," and often in years to come, he would help them out.

In Edinburgh, too, Floyd, who ever since his boyhood had had a passion for animals, saw some pedigreed Scottish terriers he wanted to take back to Kono Tayee. Two of them, extremely lively animals, were recorded in the Scottish Kennel books as

"Pepper" and "Mustard." These dogs would join his beautiful Gordon Setter pups at Clear Lake.

From Edinburgh Floyd sent to Faxon D. Atherton, member of the Second Lick Trust, a detailed account of his extensive, often conflicting findings. He found the "eminent scientists" so divided in their opinions on the comparative merits of great refractors and reflectors that he had trouble making up his own mind. He spoke of the superiority of reflectors for spectroscopic work because of their freedom from chromatic aberration. He agreed that the search for the determination of the physical constitution of heavenly bodies through spectroscopic analysis would lie in the "largest, if not the only remaining field of astronomy which now contains undiscovered truths within reach of telescopes."[37] But the problems of flexure in the mounting, of sensitivity to atmospheric disturbances in an open-tube telescope, seemed insurmountable. The general feeling was that the largest glass mirror that could be successfully made was about 4.5 feet in diameter. Therefore a silvered-glass reflector as the most powerful telescope ever constructed seemed at this time out of the question.

He had little doubt that in the current state of the art a refractor up to forty inches in diameter could be built. It was principally a question of making the glass disks successfully. This remained the crucial problem. Feil was confident he could make flint and crown disks 1.0 meter in diameter for $2,000 and 1.10 meters for $4,000. Floyd favored the smaller size. He saw no difficulty with the engineering for a dome and tube of large size for such a "grand instrument." They would "present difficulties only in ordinary mechanics quite easily mastered."

Floyd's letter to Atherton was dated August 24. By the time it reached San Francisco, Atherton was no longer a Lick trustee. Floyd learned afterward that on Sunday, August 27, Lick had sent a bellhop to Charles Plum, carpetmaker and upholsterer for the Lick House. He had asked him to come at once. George Schönewald, manager of the Lick House, had told Plum that Lick wanted to see him in his room. Plum had rushed upstairs and found the eighty-year-old man in bed, his face drawn, his eyes watery.[38]

Lick said, "Good morning, Mr. Plum." Plum replied, "Good morning, Mr. Lick. Mr. Schönewald tells me that you have sent for me."

"Yes," said Lick. "I have sent for you to give me a new set of trustees."

"A new set," Plum exclaimed. "Why surely Mr. Lick you are not going to change your trustees again." Using a violent expression, the old man said, "I *have* changed them." He showed Plum a document drawn up by his attorney, deposing four of his trustees.

Plum, afraid any new arrangement might be only temporary, hesitated, then pleaded with Lick. "*Don't* change your Trust." Lick was insistent. "Mr. Plum, do you wish to give me a new set of trustees, or shall I call some one else?" Plum, knowing Lick well, realized that to fight his wishes would "ruin the whole affair." He said he could not decide at once but would return at 3 P.M. to give him an answer.

It was then 9 A.M. Lick noted the time on his watch beside his bed. Plum left. He went first to find his attorney, Edwin B. Mastick, who lived in Alameda, across the bay. Mastick said immediately that it would be disastrous to do anything of the sort. If possible, Lick must be prevented from carrying out his intentions. They hurried down to the Alameda wharf, but the boat was delayed and they did not reach Lick until the stroke of 3. They found him, watch in hand, waiting. When he saw them, a look of satisfaction crossed his face and he lay back on his pillow with a sigh of relief. He evidently believed that Plum was going to carry out his wishes.

Plum then said, "Mr. Lick, allow me to introduce to you Mr. Mastick." Lick replied, "I know Mr. Mastick very well." Mastick spoke: "Surely Mr. Lick, you are not going to change your Trust again." Said Lick, "I have already done so," and he showed him the document deposing the trustees. He asked then, "Do you wish to become one of my trustees?" Mastick answered, "This being Sunday, no legal business can be transacted today. We will be here tomorrow morning at nine o'clock and arrange the matter." [39]

The next morning Plum and Mastick called again. For trustees, they proposed the names of Plum, Mastick, William Sherman, and Schönewald.[40] Lick quickly accepted those names. Floyd remained president, Sherman became vice-president, with Mastick as treasurer.

Lick's change of heart created a difficult situation, especially with the able trustees of the Second Trust, leaders of San Francisco society, who had no idea why they were being fired, or why Lick had changed his mind. Some said afterward that Lick had been angry because the trustees would not let him dismiss his illegitimate son John from that board. Others said that John, as a member of the board, had refused to sign both the Trust deed and the contract for the Fredericksburg monument honoring James Lick's grandfather who had fought in the Revolution under George Washington.[41] James Lick was so furious that he decided to dismiss all the trustees, with the exception of Captain Floyd, who was, of course, in Europe. Mathews, who remained as secretary, said that for a while there was "utter confusion."[42] The second board could do nothing legally. The third board could not act until each trustee was formally elected by ballot as a successor to a retiring trustee. This took nearly three months. On November 29, 1876, the court confirmed the new board.

Meanwhile, to save the Lick properties, Mathews, on whom most of the Trust burden fell, paid $40,000 in taxes, which he had managed to scrape together from rents on Lick property. In December he turned this money over to the third board.[43]

At this point many people questioned Lick's sanity. Some newspapers demanded that the trustees return all Lick's money to the old man. Everything seemed to be thrown into question. Lick finally insisted that a panel of eight prominent physicians examine him. The doctors found him sane.[44]

But Lick was feeble and tired. Each day, while Floyd was in Europe, Mathews had come to see him and to read extracts from the Captain's reports and to discuss other matters. One day Lick told him, "I have placed these matters in your hands and I don't wish to be consulted on them again."[45]

The Lick trustees, chosen in this hasty, rather haphazard way, were an interesting, hardheaded group who would serve Lick selflessly and well through the difficult years ahead. Like Floyd and Fraser, they were determined to carry out Lick's plans for the observatory, as well as for his other benefactions. The task would prove a much greater ordeal than any of them dreamed.

Looking back long afterward, Charles Plum noted that Lick would be remembered always as a promoter of science and art, a benefactor of his species. But he questioned whether he and the other men who so ably had devoted their invaluable time to carrying out the provisions of the Trust, at great sacrifice to themselves and their families, were not entitled to the gratitude of the community as much as the man who bequeathed his fortune to the public, when in the presence of death, he found it slipping from his grasp.

Plum, to whom Lick had turned for the formation of his Third Trust, was a native of New York City.[46] He had had only an elementary school education, after which he had trained for five years as an upholsterer's apprentice. He had come to California as a forty-niner. He was an organizer of the Tiger Fire Engine Company and an active volunteer member. During the "Rebellion" he had been a lieutenant in the First California Dragoons. He had first met Lick when the Lick House was being built. The two met again when Plum was elected president of the Mechanics Institute.

In 1870 Plum, a man of "great energy and enterprise," moved his business next door to Lick, and they became more intimate. Plum soon learned that Lick insisted on his own way in everything and brooked no interference in his personal affairs.[47] He became acutely aware of this on a day in 1875 when he visited the old man at Calistoga. Lick spoke of his fortune and its disposal. Plum asked, "Mr. Lick, why do you wait until you are dead? Why don't you carry out all these things while you are alive?" Lick flew into a rage. He sat up in bed and exclaimed, "Young men are always visionary; you are always interfering with my affairs." After that Plum did not see Lick again until the message came to call at the Lick House. When he felt that death was near, Lick had probably recalled Plum's words of warning.

Another trustee, George Schönewald, an old friend of Lick's, was born in Westphalia in Prussia. As a youth he deserted from the army and escaped to Holland. He arrived in San Francisco in 1866 and worked in a bakery until he found a job at White Sulphur Springs near St. Helena. In 1874 he leased and became proprietor of Sam Brannan's hotel at Calistoga Hot Springs resort and first met Lick, who had gone there for treatment.[48] He was kind to the old man, and in 1876 Lick asked him to become maître d' at the Lick House.[49] In the difficult years ahead Schönewald, with his connections in the hotel business, would be of great help to Floyd and the Lick trustees when they were faced with adverse criticism. After the Lick House he would go to the Palace Hotel in San Francisco and the Del Monte Hotel in Monterey. In the Napa Valley he became a pioneer viticulturist.

Edwin Mastick, the treasurer, an "industrious and scrupulously honest" attorney, had a library of books on astronomy and would take the deepest interest in the building of the observatory.[50] The vice-president, William Sherman, who had been persuaded to join the Trust by Plum and Mastick, hailed from Rhode Island. In San Francisco he had been for many years in the luggage and clothing business. In 1873 he was appointed subtreasurer of the United States for San Francisco. He had served also as chairman of the Republican State Central Committee. He would die before the observatory was finished.[51]

Meanwhile, Floyd, after a quick trip from Scotland to Italy, had returned to Paris on January 3, 1877, this time for a longer stay. He loved the "City of Light," a city still gay despite the recent war, and Cora shared his love. Fortunately both spoke French well. They visited the Louvre, the new, highly ornate Grand Opera House, and the Comédie Française. During the day, Cora might take the pudgy, imaginative Harry to watch a balloon ascension at the Hippodrome or to see the amusing menagerie in the Jardin des Plantes. They spent hours in the Luxembourg Gardens, where, in the distance, they could see the Paris Observatory dome. There, at day's end, Floyd would join them. At night, with Harry tucked in bed, the two adults might go to a bistro in Montmartre or to one of the glittering Cafés

Chantants. It was in a way a second honeymoon, more glamorous and different in every way from the first at Clear Lake.

Yet Floyd continued to be plagued by his sense of responsibility for the telescope, his concern over the other instruments for the observatory. At the Paris Observatory, he conferred with Leverrier and the Henry brothers, Paul and Prosper; Leverrier even invited him to dinner and a meeting at the sedate Académie des Sciences. When, some months later, after his return home, Floyd learned of Leverrier's illness and death,[52] he was "saddened at the loss of advice and counsel of a great mathematician and savant, of the most distinguished international fame."[53] He was grieved "to part with the hope that I have entertained of some day having the pleasure and honor of entertaining him in California."

At the Académie, too, he again saw Jules Janssen, who, in 1875, at Meudon outside Paris, had founded an observatory for spectroscopic work and solar photography. A man of about fifty, he had been lame since childhood. Yet, in 1870, during the siege of Paris, he had escaped by balloon from the beleaguered city to Algeria, carrying his reflector designed for the study of the sun's corona. He had reached Oran just in time for the eclipse, only to find himself "shut behind a cloud curtain more impervious than the Prussian lines."[54]

To the Floyds in Paris, signs of that siege, as results of which Napoleon III had been overthrown and Paris had fallen under German rule, were visible everywhere. There were holes in the ground where houses had been, and holes, made by shot and shell, in the Arc de Triomphe, even in the Paris Observatory. Still, the destruction caused by the Germans could not compare with that caused by the Communards, who, in their efforts to wrest control of the city from the established French government, had burned and gutted the beautiful Hôtel de Ville and the Tuileries, killing many citizens in the fierce struggles along the Rue de la Paix and the Place Vendôme. The observatory, with its massive stone walls, had been used by the Communards as a fortress. The dome had been riddled with bullets. Not until May 28, 1871, was order restored by government troops. Now, six years had passed, and, as the Floyds soon found, the psychological wounds among the people were still fresh.

Again and again during those days Floyd returned to the Rue le Brun to see the Feils. From Charles Feil he learned how the glassmaker had inherited the business from his father-in-law, Pierre Louis Guinand, a Swiss bellmaker from the canton of Neuchâtel. After countless failures Guinand had succeeded in making some perfect 4- and 6-inch crystal disks. He had taken a clay stirring rod, and operating it by hand, he developed an ingenious method of removing the striated portions of the glass that had caused him such trouble, thereby producing a homogeneous 18-inch glass. He took this disk to Paris, where it was bought by Cauchoix. A royal invitation to settle there followed. But it was too late. Guinand died the following year. His son Louis, taking his father's secret with him, moved to Paris. There he divulged it to Arno Bontemps, director of the glassworks at Choisy-le-Roi, who had fled from France during the 1848 Revolution and carried the secret to Chance in Birmingham, England. In Paris, Feil took over from Guinand and the business stayed in the family. Thus, as Feil told it, the revolution in the optical industry and the growth of telescope making came about.

Now, in Paris, Floyd visited the foundry, with its roaring furnaces, huge clay pots, holding from 500 pounds to a ton, into which the glass was poured; he also saw its annealing ovens, which had a mysterious air. Like ancient alchemists, the workmen, their eyes shielded from the fierce heat and light, mixed the batches for the glass until it became molten in the furnace. When it was entirely melted they inserted an iron stirring rod. Hour after hour they continued to stir, going faster and faster until the heat was turned off. The long, slow process of cooling then began. For a month, at least, the entire mass, pot and all, was placed in the annealing furnace. When it was taken out, the pot and the outside parts of the glass were broken off to see if the lump suitable for the required disk could be found at the interior.

One day, when Floyd happened to be at the foundry, Howard Grubb arrived from Dublin to inspect the crown disk for the

Vienna refractor. He talked glowingly of the future of refractors but favored also an 8-foot silvered glass reflector or one of speculum metal. He had received news from Melbourne, Australia, that Thomas Grubb's 4-foot speculum reflector, once considered a failure, had "by the accidental turning of some screw for another purpose, suddenly got over its imperfect figure." It began to work beautifully. For this Grubb was a "little elated." Some time later Trustee Plum would journey to Melbourne to see the result.

Grubb made a collection of photographs for Floyd and wrote out a statement on the advantages and disadvantages of different forms of telescopes. He even suggested that a reflecting telescope might be placed at the bottom of a great well, excavated in the rock, and closed by a lid to the level of the summit of Mount Hamilton. He would also make a plaster model of the mountain top, based on topographic notes Floyd had given him in Glasgow.

Newcomb had proposed a silvered-glass reflector seven feet in diameter of the Cassegrain style as preferable to the largest refractor. Floyd still thought this "almost impossible" for the Lick trustees, "as we can get nobody to make the glass except Mr. Grubb, who if we should so desire, would undertake the experiment of making the glass himself." This, at least, was the proposal Grubb had made to Newcomb. Floyd was incensed. He thought the idea manifestly absurd. They would have to pay for a glass factory to educate Mr. Grubb in the art of glassmaking, "indulging in expensive experiments with all the chances against success, and the positive certainty of an incalculable expense."[55]

Of Feil Grubb said, "He is very prolific in promises and not so clear on keeping them." For this reason he hesitated on the building of a refractor larger than the Vienna 27-inch.[56]

By this time Floyd was more than ever convinced that a large refractor as the principal instrument for the Lick telescope was the best choice. Still he hoped they might have, as a subsidiary instrument, a reflector of 40 or 50 inches in diameter with metallic and silvered-glass mirrors. With the two mirrors it would cost about $20,000. "A complete observatory in these days must have both."[57] If Newcomb heard of Floyd's latest proposal, he doubtless thought the Lick president was suffering delusions of grandeur.

4

Transition to the skies over Mount Hamilton

On October 1, 1876, Floyd received a cable from George Schönewald in San Francisco informing him that in the early morning of that day James Lick had died at the age of eighty.[1] Floyd wrote immediately to Mathews, assuring him of his confidence in his "discretion and integrity, ability and carefulness. The fact that you are the Secretary of the 'Lick Trust' under present extraordinary and complicated circumstances, prevents me from feelings of grave anxiety."[2] He promised to do all in his power "as a gentleman and as a trustee, to effect the objects of Mr. Lick's Deed of Trust."[3] At this point he had no idea how hard this would be.

In the weeks to come he learned more of Lick's death from Mathews's letters and from clippings on the elaborate funeral ceremonies in San Francisco. He thought sadly of the frugal old man, demeaned in life, now glorified in death with all the city's honors. For three days flags flew at half mast while Lick's body lay in state in the hall of the Society of California Pioneers. For three days those who had known him, and thousands who had not, filed past the coffin to gaze on the emaciated face of the "great philanthropist." Before the coming of the end he had so long feared, Mathews, who had seen him nearly every day since January, said Lick never smiled. He reflected, "I don't suppose his money has ever made him happy."[4]

On October 4 the funeral procession, bearing his coffin on a hearse drawn by four black horses, moved from Pioneer Hall, followed by three hundred members of the Society of California Pioneers and the Lick trustees (with the exception of Floyd), then Mayor Bryant, city and federal officials, the state militia, and the University of California regents. The cortege moved slowly down Montgomery to Market Street, through Sixth and Mission to the Mechanics Pavilion where the funeral service was held. Burial was in the Masonic vault. There the body would lie until, more than ten years later, it would be "borne aloft to Mount Hamilton," and Lick would have his final wish to lie under his great telescope.[5]

Just before leaving for Europe Floyd, as he told it long afterwards, had gone to see the old man. He found him highly agitated. "What," he demanded, "will happen to my body when I die?"[6] He had suggested to the trustees that a specific sum be set aside to provide for a proper tomb. But to one who had had such grandiose dreams for monuments to his parents and to himself, any ordinary tomb somehow seemed inadequate and nothing had been arranged. When Davidson suggested that he be cremated, Lick replied emphatically, "No, sir! I intend to rot like a gentleman!"[7] It was then that Floyd, thinking of Lick's greatest monument on Mount Hamilton, suggested to the old man that the Lick Observatory would be his most appropriate burial place, and the base of the pier of the great telescope, the most fitting tomb for his body.[8] Lick was fascinated by the idea, and Atherton and the other trustees quickly supported it. "This," Floyd wrote, "relieved him of much anxiety that he had been feeling about having neglected to set aside a specific sum to provide a tomb for his body – and he thereupon withdrew his request to the Trustees to set aside a specific sum for that purpose." When Floyd took his final leave, Lick, realizing perhaps that he might not be there when Floyd returned, begged him over and over to see that he was entombed under the great telescope on Mount Hamilton.

After Lick's death, Cora, tired of telescopes and observatories, insisted on a rest on the Riviera. In Nice, while they reveled in the glorious weather, swimming and basking in the sun, Dick borrowed time to thank Mathews for a flood of letters.

From the observatory in Marseilles where he saw the first silver-on-glass reflector, invented by Foucault, he journeyed on to Austria and Germany, making visits to Johann Palisa in Vienna, Anton Krueger in Kiel, and Arthur Auwers in Berlin. In Munich Sigmund Merz had said that his firm was not prepared to make the glass disks, claiming it was still necessary to know if it was possible to make homogeneous lenses of such size as 3-foot aperture "by me or other manufacturers." [9]

Everywhere he went Floyd kept detailed accounts of his expenditures and those of his wife. In Italy they traveled from Rome to Naples to Venice, enjoying the sight-seeing while picking up treasures for Kono Tayee. The bills they saved show expenditures in Paris and London, Florence and Milan, Naples and Venice, Zurich and Geneva. They cover hotels, meals and services, Venetian glass, silver and jewelry, even marble-topped tables, with vintage wines shipped around the Horn.[10] Later Floyd would be condemned by some who called his European trip a junket at the expense of the Lick Trust. Actually everything was at his own (or Cora's) expense.

On November 1, just a month after Lick's death, his son John Lick arrived in San Francisco. He demanded $385,000 for himself and his Fredericksburg relatives in addition to the $150,000 specified by Lick in his last deed. Soon his cousin Jimmy, who had also worked for Lick, turned up to protest John's application for letters of administration on his father's estate; he claimed his "illegitimate" cousin had no right to them. News of the squabble between the cousins, who had shown little concern for James Lick while he lived but wanted to share his wealth after death, had reached Floyd in Nice. He had then opposed any compromise between John Lick and the Lick trustees. Now, faced by John's demands, his threats against the Trust to prove his father "incompetent to execute the same," Floyd saw the need for "compromise."[11] A newspaperman agreed: "If the compromise is not made John H. Lick may get the whole estate and away go telescope, donations and all." John Lick had hired the able Hall McAllister as his attorney. On January 19, 1877, to avoid long and expensive litigation, the Trust agreed to a settlement of $535,000, with the proviso that all family claims be settled with this sum.

Meanwhile, discouraging letters about the Lick estate continued to reach Floyd from Mathews. Afraid the problems might delay work on the telescope for months, he thought it unwise to make any contracts for the great telescope until they were settled. He noted that "constructors of work requiring *so much time demand tremendous margins for uncertainty.*"[12] The delays were agonizing. Hoping for an early settlement, Floyd told Mathews he would be happy if, before leaving Europe, he could help the Third Board of Trustees in any way. He suggested that Grubb and Clark be asked for bids. They were the only men he would trust to figure the large disks.

The Floyds finally sailed for New York in June, 1877, after thirteen months away from home. The crossing was rough, and Cora was so prostrated by seasickness that they were forced to stay in New York until she recovered. It was a great day when they arrived back in San Francisco, a city that had changed radically since Floyd had landed there after the Civil War. As it had spread out into the harbor, paved streets now ran where ships once had anchored.[13] As the crowds on those streets and the coaches and wagons had increased, the buildings had grown "to produce the effect of a suddenly risen city of enchantment."[14] Even the Lick House now had a rival in the magnificent Palace Hotel. In July, 1877, the dining room of the Lick House was destroyed by fire. The Lick trustees tried to restore it as nearly as possible to its former state, but it was never the same.

Floyd, delighted to be home, was disturbed to learn that controversy continued to rage in the wake of Lick's death. As in the two previous deeds, the telescope was only one of Lick's bequests. These included the monument to his German grandfather, who had fought in the American Revolution; diverse amounts for a Protestant orphan asylum in San Francisco, another in San Jose "for all orphans, without regard to religion of parents"; the Ladies Protection and Relief Society of San Francisco; the erection and maintenance of free baths in San

Francisco–an institution that would provide for fifty thousand baths, surpassing all other benefactions except the telescope; a bronze monument to Francis Scott Key in Golden Gate Park and bronze statuary portraying the history of California at the San Francisco City Hall; and the founding and endowment of "the California School of Mechanical Arts."

The residuary legatees (to share equally) were the California Academy of Sciences and the Society of California Pioneers. To Floyd it seemed obvious that the $535,000 awarded to John Lick should come out of the coffers of these legatees instead of being divided proportionately among all the beneficiaries. This remained the trustees' view. The legatees disagreed, and the case finally went to court. Not until January 14, 1878, was it decided: The residuary legatees must sustain any loss "no matter from what cause or accident it might occur." [15] John Lick's demands had to be paid out of their pockets. Davidson was more bitter than ever.

Even then one serious problem remained. According to the state, Lick had not paid some 1868 taxes on land he owned in San Francisco. He had insisted that any further payment would mean double taxation and would be unconstitutional. Nevertheless liens remained against his estate for those taxes.

In a statement to the Senate Finance Committee in Sacramento, Floyd argued for the Lick trustees that Lick could hardly have failed to pay his 1868 tax without good reason; that the taxes he had paid over a period of twenty-eight years must be enormous; that, in any case, his vast estate, donated for the benefit of "the whole people," should be "free from taxation." [16] James Lick, he pleaded eloquently, had begun the battle of life a poor boy, had donated almost his entire fortune to the advancement of science, the relief, elevation, and improvement of his fellow creatures. He was indeed "one of the greatest benefactors of his country and his race." Despite such arguments, the case dragged on for months before it was finally settled in the trustees' favor. Only then, with all the legal problems resolved, could the trustees go ahead with the sale of the Lick properties to obtain funds for specific bequests. This proved to be a huge, often thankless task.

Soon after Lick's death his meager personal property at the Lick homestead had been sold. In addition to horses, colts, cows, and his cart, there was one marble table, one writing desk, a hand organ, and a music box. Yet, even before this, after the First Lick Trust was formed, an advertisement had appeared in the San Francisco newspapers for an auction sale at Platt's Hall on Montgomery Street near Bush. It read: "Grand Special Sale of Real Estate Per Order of the Trustees of the James Lick Trust. Situated in San Francisco, Santa Clara County, Los Angeles County and Virginia City, Nevada. San Francisco Property includes the Lick House property – southwest corner of Montgomery and Sutter Streets, Southwest corner Market and 4th Streets." [17]

At that time only enough of Lick's property was sold to pay the immediate expenses of the Lick Trust. In the years to come the rest of Lick's properties had to be sold. These ranged from the 45,820-acre Santa Catalina Island to a part of the 6,647-acre Los Felis ranch in Southern California, from the vast holdings in and around Santa Clara County and San Francisco to the Lick House. The job, Floyd realized, would leave him little time for the life of leisure he had once planned at Kono Tayee. He was also worried about Cora, who was having a difficult pregnancy and wanted him near. On November 10, 1877, baby Julia was born prematurely. She lived only one day, and her sorrowing parents buried her in Laurel Hill Cemetery next to her grandfather.[18]

During the summer of 1877 the Floyds were living at Clear Lake. One day Dick set out from Kono Tayee for Mount Hamilton, on the trek he would take so often in the years to come. That trek by "mountain road and axle" over the steep passes was long and hard. The trip to San Francisco alone took eleven hours. From Kono Tayee he sailed across the lake to the Soda Bay Hotel at the foot of Mount Konocti. There he could catch a stage connecting with the six-horse Concord Stage, running from Lakeport to Calistoga and Vallejo, or the one that ran from Lakeport to Cloverdale and Petaluma. To reach Calistoga he took W. F. Fisher's six-horse stage under the charge of that "prince of Jehus," Joe Johns ("No more careful, accommodat-

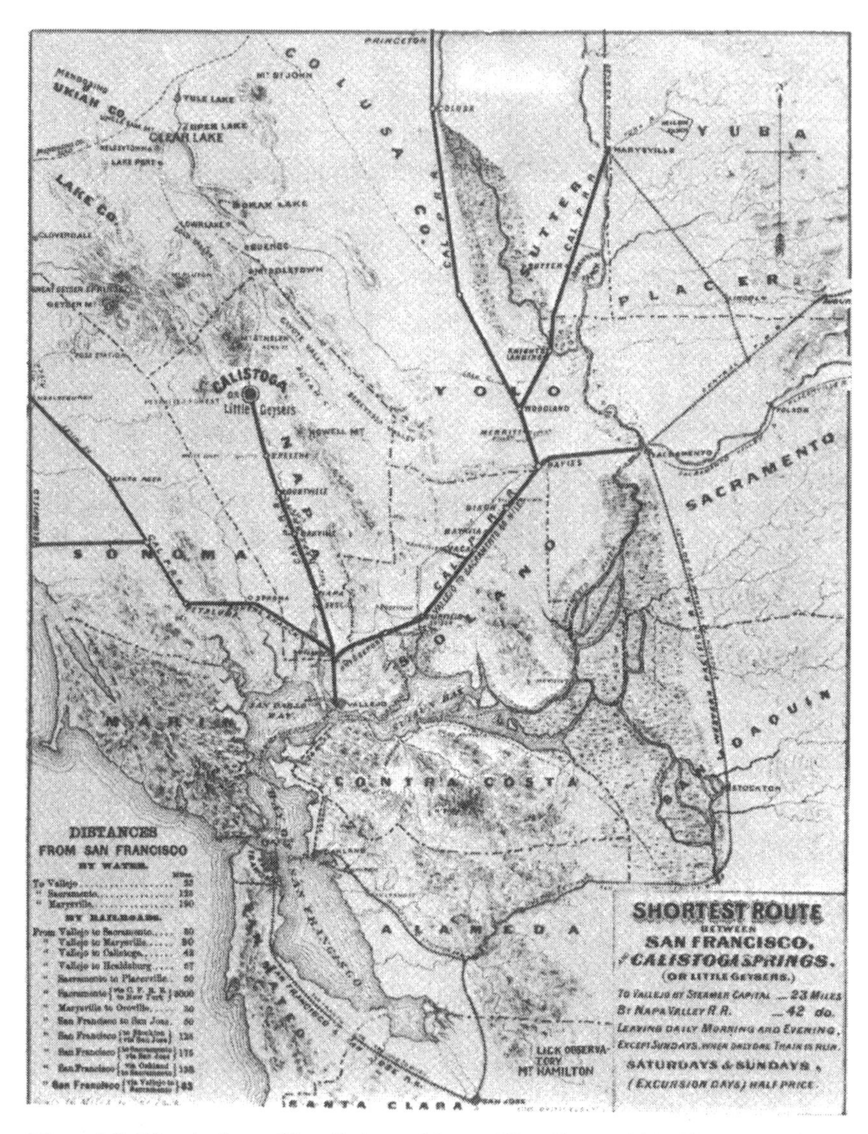

Map of California from Clear Lake to Mount Hamilton, 1871. (Courtesy of the Bancroft Library, University of California, Berkeley.)

ing, pleasant companion ever pulled a rein or whip"). At Vallejo he could then catch a ferry to the Market Street wharf in San Francisco. This route, the most direct, was described as the "most picturesque and romantic route on the Coast." From Mt.

St. Helena," the stage line proclaimed, "it affords the traveler a beautiful view of the far famed Napa and Russian River Valleys and mountains of the Coast range; and from Cobb Mountain the great Clear Lake region in front and the Pacific in the distance." This was the route Floyd usually took. The stage ran three times a week in each direction.

But he could also go by way of Cloverdale, terminus of the San Francisco and North Pacific Railroad. The stage ride, over rough roads, up and down steep, winding, precipitous mountain trails, was often wild. The driver needed the greatest skill and daring to guide and control his galloping horses. There Floyd could catch a morning train, running through the Russian River Valley to Tiburon, ninety miles away, across the bay from San Francisco, where he boarded the ferry for the city. In San Francisco, in either case, he would catch the train for San Jose where, finally, he could climb the mountain on foot or take the Mt. Hamilton Stage – an interesting, often dramatic journey, at times dusty and hot, again thoroughly wet and cold, especially in the rainy season when the roads might be washed out. Whatever route he took, it was always a long day's journey, often longer, and at times he must have dreaded it.

Mount Hamilton was a magnificent site for an observatory, but there were those, Floyd knew, who still opposed it. He decided, therefore, to invite the members of the French commission to observe the transit of Mercury of 1878 from Mount Hamilton. When they were told that observing conditions there were "foul," they had gone instead to Ogden, Utah. They were greeted by a snow storm; on Mount Hamilton the sun shone brightly.

Undaunted, the French party continued on, arriving in San Francisco in May 1878. On May 18, at 8 A.M., Floyd met them there with Tom Fraser, who, after Lick's death, had transferred his services to the Lick Trust. They boarded the train and at 11 A.M. arrived in San Jose. They climbed into the two waiting carriages, each drawn by a four-in-hand team. From the first the Frenchmen were enchanted; as the carriages climbed the mountain, they became ecstatic over the views. At Smith Creek they stopped to unpack the "bountiful lunch," prepared by S. W. Churchill of the Auzerais House in San Jose and by George Schönewald of the Lick House. They spread it on the grass under the trees overhanging the creek. A San Jose reporter, who joined the group, noted that "the savants proved themselves well up in gastronomy as well as astronomy." [19]

After lunch they began the mountain climb. As they ascended, following the many twists and turns to reach the top, they could look out over the Santa Clara Valley to the Santa Cruz mountains on the west, with Monterey Bay beyond. To the southeast the majestic Sierra Nevada rose. In the foreground, almost at their feet, it seemed, lay San Francisco Bay, with Mount Tamalpais and the entrance to the Golden Gate in the distance. On an exceptionally clear day a full-rigged ship, all sails set, could be seen through a glass, passing through the Golden Gate, entering San Francisco Bay, fifty miles away.[20] For the climb most of the party had boarded the carriages. But Philippe Hutt, an engineer in the French navy, decided to walk. Floyd and Charles André, professor of astronomy at the University of Lyons, joined him. Dressed in "a neat mountain costume," Floyd strode along, "taking one step to the Professor's six," all the time gesticulating, trying to carry on a lively conversation in French. At times, as he struggled to find the proper word, André would urge him on, crying "All right, all right" in ringing tones that echoed down the mountainside. The Captain, encouraged, would begin again, first in excellent French, then gliding into broken English "and tapering off into the broadest kind of American vernacular." Two and a half miles from the top, the road, damaged by the winter's storms, became impassable; the carriages were forced to take to the ridge, and soon the entire party was walking over the rough terrain. But they were delighted by the adventure, by the beauty of their surroundings at the summit, an island above the low-lying fog. They soon joined Floyd in a discussion of the location of buildings and instruments, and plans for the 1882 transit of Venus when he hoped the needed instruments would be ready. The reporter, listening in, noted confidently, "All who know the Captain know that if he sets his mind on this it will be done. . . . He will use his best energies to have the observations taken here, the most perfect of any taken in the world."[21]

That night Floyd set up his little $2\frac{1}{2}$-inch telescope. For an hour and a half, Alfred Angot of the École de la Fontaine in Paris, with the others, looked at diverse objects, exclaiming over the excellent images and the quality of the "seeing." At 11 P.M. some of the party turned in. Professor André stayed up with Floyd all night, "walking around the camp and hills, talking science and occasionally halting to take a peep through the telescope."

The next day the commissioners had to catch the 8 A.M. train in San Jose. Those not already awake were roused at 3; they were ready to start at 4. But Floyd had turned the telescope on Venus. André was entranced. It was 4:30 before they could be "dragged away." At 4:45 the party took off. They landed in the San Jose station at 7:45. "This necessitated some tolerably reckless driving down the grade, but the Commissioners kept their seats and they were brought in on time." Before leaving, Professor André asserted that at no other place in the world did they find everything so highly favorable for astronomical observation. The Lick Observatory instruments, he claimed, would have advantages superior to any in existence.

So ended the first of many scientific expeditions to Mount Hamilton. The reporter concluded, "We hope this will silence those who have been so energetically striving to lead the world to believe the location of the telescope here will result in a failure of the object which Mr. Lick hoped to reach by his munificent donation." Floyd and Fraser naturally agreed, but they knew it would be hard to silence those, like Edward Holden and George Davidson, who claimed that if they had been responsible, another site would have been chosen and the work would have moved faster, with more promising results.

Nevertheless, a more thorough examination of the site was obviously needed. Floyd had long wanted the expert opinion of an astronomer who could test the "seeing" over an extended period. As early as 1874, Newcomb, with Clark and Holden, had urged that the observations be made by Sherburne Wesley Burnham of Chicago, a court reporter by day, an amateur astronomer with an international reputation by night.[22]

On June 25, 1879, a year after the Frenchmen's visit, Fraser, now superintendent of construction, loaded three four-horse teams with lumber and "furnishing goods" and set out on the Mount Hamilton road to prepare for Burnham's coming. Up that road fuel, lumber, stone, lime, and all other building materials would have to be hauled in the years ahead – along with pigs, cows, and chickens to feed the mountain inhabitants. With a reasonably good team the trip up usually took four hours, much longer with loaded wagons; the return took three.

At 2 P.M. Fraser reached the top. With three carpenters and one laborer he set to work to make a shelter for the astronomer and his telescope against the wind, which, as Floyd pointed out, "blows like fury" over the summit. "No tent will stand." By night Fraser had closed in a twelve- by sixteen-foot kitchen. While the wind howled outside, he slept with his men inside. "I feel very comfortable in my new quarters," he wrote in his "log."

> The site is beautiful. Looking out of the west side of the House the whole of Santa Clara Valley and San Francisco is in sight. The sunset tonight was delightful. The poet may well exclaim:
>
> Bright is the summit range of Santa Cruz
> Shaded its slopes with many evening hues.
> The glistening bay reflects the town of oaks
> And o'er San Bruno rise Frisco's smokes.
> And yonder gleaming in the sun's last ray
> The ocean shines off Monterey's calm bay.
> Forming the fit surroundings of the throne
> Which the James Lick Trust will raise for science on this cone.[23]

Three days later, working his men late and early, paying them $3 a day, Fraser was ready for Burnham. The astronomer arrived in mid-July. A black-haired, casual-looking, wiry little man, pale and thin, he had the keenest, most penetrating eyes Floyd had ever seen. He was delighted with the arrangements on the mountain, and the longer he stayed, the more enthusiastic he became, and the more Floyd and Fraser liked him. Burnham stayed continuously from August 17 to October 16. During that time he had forty-two "first class" nights and seven medium ones; eleven were cloudy and foggy. Working with his excellent 6-inch Clark refractor he felt that there was only one class of objects by which he could judge the atmospheric conditions on the mountain or could thoroughly test an object glass or mirror. This was by discovering, observing, and measuring difficult double stars, particularly those that are both close and unequal. Such objects were his specialty.

For the first time on Mount Hamilton Burnham was able to observe the southern zones of the sky in the region lying more than 30 or 35 degrees south of the celestial equator. "That to me was a new heavens *[sic]*, and the most promising field for such research." He discovered that he could observe close double stars down to forty-three degrees south declination. He was impressed by the "wonderful purity and steadiness of the air almost down to the very horizon." Even with a thirty-mile-an-hour wind, the "seeing" did not appear to be affected.[24]

Night after night, at sunset, Burnham watched as the ocean fog 2,000 feet below rolled in from the Pacific at the Golden Gate on the north and Monterey Bay on the south, covering the whole valley between the base of Mount Hamilton and the Coast Range with a dense mass of vapor that he compared to a great white sea, with the tops of the lower hills standing up through it like islands. Above this fog layer, the skies were clear.

After a month and a half on the mountain, Burnham took time out to scribble a note to Floyd in San Francisco, "Everything is going on swimmingly and I have been running at high pressure. There is more weather here to the square mile than any place I have ever heard of."[25] Working "about all night lately," he had been getting some good observations. During the day he had so much tinkering to do that he had not spent much time in sleep. Even by daylight he had made some measurements – notably of the sixth star in the Orion trapezium – to see what could be done there. In addition to various new doubles, he had discovered some "specially interesting tough customers." "This place," he exclaimed, "will stand investigations of this kind." His only complaint was that he had little time for photography and none for hunting. (He was an excellent shot and an artistic photographer.) "But," he noted hopefully, "that may come yet. The wild beasts of the field, and the fowls of the air haven't manifested any impatience at being let alone, and will probably continue to live if this monotonously good weather continues."

Before he left California, Burnham accepted Floyd's invitation for a week's shooting with Tom Fraser at Clear Lake. The chase after big bucks on Mount Konocti was successful, and the entertainment, food, and drink at Kono Tayee were first rate.

After his return to Chicago, Burnham sent a detailed report of his observations. "There can be no doubt," he said, "that Mount Hamilton offers advantages superior to those found at any point where a permanent observatory has been established. The remarkable steadiness of the air, and the continued succession of nights of almost perfect condition, are conditions not to be hoped for in any place with which I am acquainted." Looking back nostalgically on his expedition he wrote, "The saddest picture of all is to be found at this end of the line. I am ruined by cheap Chinese labor. I don't think I shall ever again smile out loud. Here I have been 5 weeks, and I will be hanged if there has been even a third class night in the whole time. The poorest part of the poorest night I ever saw in Cala. was a great deal better than the best part of the best night here during this time. Now, if I had stopped at home, I should have thought it was all right, but having tasted of the tree of knowledge, my astronomical eyes have been opened, and I am ashamed of the weather on this side of the hemisphere." [26]

He warned Floyd, "Don't get many observers out in Cala. for it will ruin the business everywhere else. The fellow will be so disgusted when he gets back home that he won't be good for anything – for a while at least. I don't see any other way than to annex all the States so that Cala. weather will come clear across."

Everywhere Burnham found people enthusiastic over the new observatory. "You will find," he wrote to Floyd, "that the building of this Obsy. will attract more attention and interest than anything of the kind ever done before." He hoped work could begin soon. "Every season that a large glass there is not at work, is an immense loss to science, and I hope you yourself will see some big things that haven't been seen, before you turn it over to anyone else. It will be an easy thing to do."

In September 1879, after a long and tiresome railroad journey, Simon Newcomb, who was now in charge of the Nautical Almanac office,[27] arrived at the Palace Hotel in San Francisco. Floyd, at Kono Tayee, had asked Fraser to meet the astronomer. "I don't know how the deuce I am to meet you on the boat as the 17th comes on Wednesday when there's no stage down by Lower Lake. However I'll try to wiggle down some way."[28] The "wiggle" worked, although it disrupted Cora's plans and caused inconvenience and embarrassment to Floyd, who was compelled to ask a number of invited guests to defer their visits until he got home.

Eager to make his guests comfortable, even at the outpost on Mount Hamilton, Floyd told Fraser to take a cook up the mountain, preferably one from the Auzerais House in San Jose, "or anywhere else you please so he is a cook. Send him up with old Coffee day before if possible and let him take a market basket along with garlic onions celery something for salad and other vegetables – fresh meat – in other words *something to eat.* I'll pay for the cook and the Trust can pay for the provender. Of course I only want him on the mountain as long as we are there."[29]

Newcomb's visit was a huge success. He spent three nights on the mountain with Floyd and Burnham, "studying the skies by night and prospecting around in the daytime to see whether the mountain top or some point in the neighboring plateau offered the best location for the observatory." As for the atmospheric conditions, the results were beyond his most "sanguine" expectations.[30] Back in Washington, after a meeting with the Lick trustees in San Francisco, Newcomb spoke enthusiastically of the remarkable "seeing." To Floyd he wrote, "I think that you can proceed with the construction of the great telescope in the confidence that it will be placed in a situation more favorable for using it than any other in which a telescope has been tried."[31] Floyd was, of course, immensely relieved. For Newcomb the visit apparently eliminated any fears of the Captain's ability to direct the building of a great observatory. He wrote warmly, "Although R.R. travelling is tedious to me, I shall always remember with pleasure my visit to San Francisco and the three nights on Mt. H., a pleasure largely due to yourself. I assure you that I feel your kindness and attention more deeply than I have ever sought to express."[32]

5

"Dear Captain"

Months dragged into four long years before the legal problems of the Lick Trust were finally settled. Floyd continued to be "perplexed and anxious" about the kind of telescope most suitable for the observatory. He told Simon Newcomb, "I have never yet been able to get down 'off the fence' entirely, and I hope you will help me down on the best side." [1]

It was a vexing problem. Feil, in Paris, argued that the "Savants" over his way considered the refractor the telescope of the future.[2] In England George Calver had just completed a 27½-inch silvered-glass reflector for Andrew A. Common at Ealing and called it superior to any large refractor. Floyd thought it must be superior to the Paris reflector, as Calver had used thicker glass and the mounting was finer. But he thought, too, that adequate tests had not yet been made. He concluded that the advocates of reflectors were too sanguine, especially proponents of mirrors more than four feet in diameter. Calver offered to send out one of his speculum mirrors. Floyd liked the idea but did not see how it could be done "without making embarrassing circumstances." [3]

At the end of July 1879, Lick trustee Charles Plum visited Howard Grubb in Dublin. He learned that Grubb, as a result of his difficulties with Feil over the Vienna refractor, thought the chance of obtaining a pair of 36-inch disks remote. Year after year might pass, he said, and the trustees die off one by one without seing *anything done* as the fruit of their labor. He feared

Kono Tayee at Clear Lake, from *Lake County, California,* illustrated and described, showing its advantages for homes by W. W. Elliott (Oakland, 1885). (Reproduced by kind permission of the Mary Lea Shane Archives of the Lick Observatory.)

the trustees might be put in a very unpleasant position. They might have to wait twenty years for the great refractor, whereas, with a great reflector, they would soon have something to show.[4]

In April 1880, Floyd left with Tom Fraser for Washington to discuss with Simon Newcomb building plans based on the drawings they had made of the mountaintop. They were joined on the long, dusty, transcontinental journey by Cora and seven-year-old Harry. The Floyds had considered renting a house and spending some months in Washington, but the strong-minded Cora, unwilling to be away from Kono Tayee and San Francisco so long, vetoed the idea. On the way, at the Washburn Observatory in Madison, Wisconsin, they saw James Watson, a potential candidate for the directorship of the Lick Observatory.[5] They took a carriage out to the Dearborn Observatory in Chicago and spent the night observing with Burnham and the $18\frac{1}{2}$-inch Clark refractor, once the largest telescope in the world. Said Fraser, "It was very fine indeed and a great treat to us."[6]

In Washington Floyd had his first real chance to talk with Newcomb's assistant, Edward Holden, whom he had met briefly in London in 1876. A large man with a big head and bulbous eyes that peered through wire-rim glasses, his straight, dark hair was parted in the middle and he wore a mustache and sideburns. Born in St. Louis in 1846, Holden had lost his mother and sister in the cholera epidemic of 1849. When he was nineteen his father suffered the same fate.[7] A few years after his mother's death he was sent to Cambridge, Massachusetts, to be raised by an aunt; there he entered a private school run by a cousin who was the daughter of another of his father's sisters. He found the intellectual atmosphere of the college town stimulating. Here he came to know such luminaries as Louis Agassiz, Asa Gray, and Charles Henry Davis, superintendent of the Nautical Almanac Office, as well as Holden's cousin by marriage George P. Bond, director of the Harvard College Observatory. On one unforgettable night, Bond showed him the bright blue star Vega through the "great" 15-inch refracting telescope. The young boy's interest in astronomy, thus sparked, grew at Washington University in St. Louis where he studied under the eminent mathematician and astronomer William Chauvenet before going on to West Point. During a winter spent with Chauvenet and his family in St. Paul, where the astronomer was staying for his health, Holden fell in love with Chauvenet's daughter Mary, whom he later married.[8]

At West Point the rigorous military training made an impression on Holden that was to influence his way of thinking and working for the rest of his life. In 1873 he joined the Naval Observatory staff in Washington as professor of mathematics.

Holden was, as Floyd would learn, an energetic, egotistical, versatile man with wide-ranging interests. During his Cambridge years he developed a passion for books and music. He loved to write and he wrote well, as his countless letters and publications show. (A friend called him "a gifted ink-slinger.") These range from a biography of William Herschel to a treatise entitled "The Bastion System of Fortifications, Its Defects and their Remedies"; from a study on rattlesnakes to one on Central American picture writing; and from a primer of heraldry to "The Mogul Emperors of Hindustan." His astronomical publications, equally diverse, include a monograph on the Orion nebula, with an exhaustive résumé of observations that he compared with his own, as he looked for changes in the nebula. This, like many of his other publications, reflects his passion for bibliographic detail.

Yet, if he was versatile in his work, if he could be a charming, at times amusing companion, Holden's ideas of running an observatory, stemming from his military training, eventually would lead to his downfall. His views appear quite clearly in a letter written soon after he arrived at the Naval Observatory. "I think I see very plainly that the Observatory here *might* be made a really first class working institution and that it only needs an infusion of some military order administered by someone who is an astronomer."[9] This, as it turned out, would be his outlook on any observatory he had anything to do with. It might provide order, but, carried to an extreme, it caused dissension, especially in his dealings with astronomers, who tend to be independent in their way of thinking and acting. The military order he sought did not encourage creativity.

Some years later, S. W. Burnham, after joining the Lick Observatory staff, protested Holden's rule that anything he wrote must be submitted to the director before publication. He defended his right to freedom of action – to give to the world the benefit of what he had done. "It is not a military institution and it is not necessary or desirable to have military or naval discipline. We are all supposed to be doing our best, for the love of the work we have chosen, and it is not conceivable that any abuse of entire personal freedom in the matter of writing and forwarding astronomical articles could follow on the part of any astronomer here."[10]

Often, too, Holden's arrogance and sense of self-importance got him into trouble. He could not understand why he ended at odds with his associates so often. When he was twenty-nine he wrote to his friend, Henry Draper, "Fighting people seems to be my fate."[11]

His passion for publicity also caused problems at times. On the night of September 4, 1877, soon after Asaph Hall discovered the satellites of Mars (which he named Deimos and Phobos), Holden thought he observed a third satellite.[12] On October 30, 1877, Hall wrote to E. C. Pickering at the Harvard Observatory, commenting that the object found by Holden violated not only Kepler's third law but also the simplest rules of geometry. Its existence was a mathematical impossibility.

In 1874, when D. O. Mills, a member of the First Lick Trust, had visited Washington, Newcomb suggested that Holden, his assistant on the 26-inch refractor, might be qualified one day to take charge of the Lick Observatory.[13] Gradually the trustees of the various boards came to accept the idea, and Holden did everything in his power to promote it.

After Newcomb's meeting with Mills in October 1874, Holden, in a fifteen-page report, summarized the details of the plan that Newcomb and Mills had discussed. He sent it to Mills on behalf of the Lick trustees, with the comment, "It has been the desire, so far, to indicate in a very general way, the objects to which the new observatory should devote itself in the future, in order for it to be useful in promoting science." [14] A. W. Bowman, secretary of the First Lick Trust, in acknowledging Holden's "valuable communication," wrote, "The matter is one of such vast importance to science and the world at large, the Trustees place great value on suggestions from so high a source." [15] Later Newcomb, who was deeply absorbed in his own research, noted that he had been quite willing that the junior astronomer should have as prominent a part as he chose in the report. Yet, when Holden long afterward claimed the "plan" as his own, Newcomb protested vehemently.

When the Second Lick Trust was formed, Holden, upset by the course of events, told Newcomb, "Mr. Lick's idea is to build and equip the new observatory and then turn it over to the University of California, which strikes me as a disastrous way of working."[16] He added, "Todd tells me that Mr. Lick does not like the reflector plan, nor the idea of getting Martin to come to America – in short that he has ideas set on a big refractor. He will miss in that way surely. I hope your advice can be carried out in the matter to save money for Astronomy's sake. I hate to see so much money wasted." Like Davidson, Holden had been unhappy over the site selected. He told Draper, "I doubt if things look as well for science (let alone for me) as before."[17] He even told Draper of his hope of making Léopold Trouvelot a member of his staff in the new observatory. "In this way he is quite unequalled and if the Lick affair had turned out differently he would have been a man whom I should have wanted."

At that time William Alvord, former mayor of San Francisco and a member of the First Lick Trust, hearing the news about the change in Lick's plan, had promised Holden to "use all the influence I have in your favor."[18]

In the years to come Newcomb and others continued to consider candidates who might take over the observatory when it was finished. Joseph Henry had proposed the English astronomer, Norman Lockyer,[19] and in 1875 Newcomb wrote to one of two prospects, asking if he would be willing to take on the job, warning that it involved "facing the mouth of a tolerably wild lion."[20]

In 1879, when Newcomb was on Mount Hamilton, Floyd had asked Holden to prepare preliminary observatory plans from the engineers' reports. In Washington Newcomb consulted Holden and a sketch was drawn. But, as Floyd pointed out, it was hard to draw up a plan that would fit the difficult terrain and conditions on a mountaintop Newcomb had visited once and Holden had never seen. Newcomb emphasized that his assistant "necessarily could give little advice about the location of the buildings." In matters of detail, however, Holden "did make a great many suggestions."[21]

Actually the first plan for an observatory on Mount Hamilton had been made by James Lick himself. He envisaged a circular wall, twenty feet high, that would protect the building from the gale winds that often blew there in winter. Inside the wall, at the center of a cement floor, the great telescope would rise. Outside the wall Lick wanted buildings for the "professors." On one of the two level ridges down the road he dreamed of a "neat Hotel" for "guests and strangers visiting the place." The rent from the hotel would help to pay observatory expenses."[22] (Some years later Floyd even proposed building a tavern on the mountain as a possible source of added income!)

Now, in Washington, Floyd and Fraser discussed the diverse arrangements of buildings they had made as a guide to a draftsman for "suitable plans and elevations" they could work from in California.[23] They discussed, too, the pros and cons of refracting and reflecting telescopes. At the end all agreed that, given the conditions on Mount Hamilton, a refractor would best qualify for Lick's dream of the most powerful telescope in the world.

A few days later Floyd and Fraser, in Cambridgeport, Massachusetts, found the seventy-six-year-old Alvan Clark in his small shop in the dingy building on Henry Street near the Charles River, where Floyd had visited him on his return from Europe in 1877. As engraver and miniature painter, Clark had first turned to lens grinding as a hobby.[24] Then he had made it his business. With his sons, George and Alvan Graham, he had ground the lenses for some of the greatest telescopes in the world.

Lick had written to Clark in 1873 about his plan to build a great telescope. Clark proposed that an experienced glassmaker be brought from Europe and that the finishing of all the parts be done at the observatory site. He referred to Lick's proposal for a site at 10,000 feet and pointed out that such a site would not be "convenient to the great centers" to which they had to look for trained workmen, tools, and materials. He suggested that the atmosphere be tested at the proposed site. He concluded sadly, "You will probably hear from me again, or my sons, ere long. I shall be 70 next March and what I do, I must do quickly."[25]

Seven months passed. Then Lick asked Clark for an estimate of the cost of grinding the lenses for his great telescope. Carried away by visions of Lick's vast wealth, Clark proposed what Lick considered the "exorbitant" sum of $180,000 *in gold* for a 30- or 36-inch lens for a refracting telescope.[26] Lick, outraged, refused to have anything more to do with the man. This created an impasse. Hoping to find an alternative, Mills sailed for Europe. On December 3, 1874, Newcomb followed as agent of the Lick trustees, visiting observatories and optical and instrument shops in England, France, Germany, and Russia. Now, with Lick's death, the obstacle to Clark had been removed.

Clark was eager to grind the Lick glasses, provided Feil in Paris could succeed in making crown and flint components of proper size and purity to form the 36-inch doublet for a refractor. Feil continued to assure Floyd and Clark that he would be successful.

While the litigation on the Lick estate went on, Alvan Clark and Sons had produced for the Imperial Observatory at Pulkowa, Russia, a 30-inch glass, the largest lens in the world. To fulfill the conditions of Lick's gift, the trustees needed a larger glass than this. The Clarks finally agreed to supply a 36-inch lens for $50,000. In December 1880, Floyd, acting for the Lick trustees, signed the contract.

When Grubb learned the contract had gone to Clark he was shocked. "Because," as he wrote to Floyd, "it will be within your recollection that I took a large amount of trouble and interest in the plans and arrangements for the Lick Observatory, the only outcome of which was that the Trustees proceeded to order an objective from Mr. Clark without even asking tenders from my house and I must frankly confess that I consider the Trustees did not act toward me with that courtesy which I think I had a right to expect at their hands."[27]

In the meantime, while Floyd was again in California, Fraser spent five weeks in Washington, working on the observatory plans in Newcomb's office, reducing their ideas to a plan for Mount Hamilton that could then be worked out by a draftsman.[28] Fraser returned home in June 1880, "bright as a new dollar," delighted to feel the California air again, and eager to get to work. On July 20 he arrived on Mount Hamilton with three men, including a carpenter, and started to build a boardinghouse, twenty-six by thirty-six feet, for the workmen. A blacksmith shop, carriage house, and hayshed followed with a "house for myself to live in" and one for maps and books, set apart in case of fire. He repaired the stable built the previous year and put it in order for the winter. In his "log" he wrote, "Got to Mt. Hamilton today. Had 2 horses in large wagon with Wm. Bent, Carpenter, Dave Graham, workman and Ben Hanahan driver. I also had my horse Cate in the buggie bringing my wife and myself, both carriages loaded to the gards." [29]

At first Fraser and his charming, efficient wife, Floretta McClellan of San Jose, whom he had married in 1875, moved into the little hut built for Burnham in 1879. Later, when the superintendent's house was finished, they moved there. Floretta (or Etta, as her husband called her) had known Tom ever since her school days at Mills Seminary in Oakland.[30] She was a gentle, kind-hearted woman who would accept the hardships of the pioneer life on Mount Hamilton, making the best of them. Through all their years she was to prove a constant help to her hard-working husband. They had no children, but, as a biographer wrote, "The wealth of affection in the great heart of the captain was concentrated on his wife. He literally worshipped her." [31] As visitors arrived in increasing numbers, he became known for his geniality as a host, and she for her warm welcome to all who came. The job was not easy. The problem of keeping supplies on the mountain was great. In the summer of 1883 Floretta Fraser pleaded with Trustee Sherman for a "barrell of butter, 2 boxes of codfish, 1 box of oysters. Have not more than enough butter to last us 6 days." She added, "The plasterers are here now and Mr. Fraser neither takes time to eat or drink." [32]

In August Floyd, anxious to see how his directions were being carried out, joined Fraser on an expedition to the "hill." They left San Jose at 4:40 A.M., had breakfast at Smith Creek at 8:40, and started up the mountain at 11. In his "log" Fraser recorded, "Captain drove horses all the way." [33] That afternoon they began to plan for further work. In his report to the trustees Fraser wrote, "Captain R S Floyd has been up twice since he returned from the East the last time he stoped 3 weeks but one day and it was a great help to me to have him to consult about the work to be done and to get his orders and Instructions right on the ground where he could see for himself what was needed to be done in starting a work of such great magnitude as the one before you and the first work at the head of the James Lick bequests." He signed his report, "I remain yours to command, Thos. E. Fraser." [34] Afterwards he announced, "We must have the best here and I will never submit to anything else while I am in charge."

A fundamental problem was the lack of a convenient water source. At the beginning water had to be hauled all the way up from Smith Creek, seven miles away, 2,000 feet below the top. A spring was then found on the roadside below China Camp; this still meant a total of six miles for a man and horse-drawn wagon to travel each day, and the water was fit to drink only on the day it was drawn from the spring. Then, on a memorable day in 1878, Fraser, exploring the wilderness area below Observatory Peak, had come on a trickle of water. He traced it to a spring bubbling with fine, clear water. Overjoyed, he could not wait to tell the Captain. He named it the R. S. Spring. But it was hard to reach. To make a trail from the peak, they had to blast through solid rock. For a time the 2,000 gallons proved adequate. But a larger supply, piped to the site, was needed. On Middle Peak Fraser found two places where he could put two 5,000-gallon tanks, but could not decide which to choose. Overwhelmed by the responsibility, he wrote in his "log," using his unique spelling and lack of punctuation, "Our whole object and my positive instructions is to doe things once and then to know that it is the best that could be don under the circumstances." [35] To which he adds, "The above is what gives us so much study and worry." He knew their work in this "great undertaking" would be closely scrutinized and critically examined by men of experience in civil engineering and mathematics. He predicted that what they were doing might not suit in all respects the scientists who would come there. But they could not stop to worry about that now. After all they had already consulted some of the greatest astronomers here and abroad. They could only do "the best we knew" and let the future take care of the final results.

Fraser finally decided to put four tanks at a high point east of Observatory Peak. He installed a steam boiler and pump at the spring. On October 22, 1880, he started to pump and in ten minutes water was running into the tanks on the hill. "Now," he exclaimed gleefully, "I got water on the top of Mt. Hamilton at last. It is a fine thing." [36] By this time, too, he was hard at work blasting the top off Observatory Peak, so that the large dome could rest on a solid foundation of natural rock. He wrote, "It is a heavy job, the highest point to be cut down is 32 feet." This, the western and lowest of the three peaks on Mount Hamilton, 4,200 feet in height, lies 2,500 feet from the middle peak, later called Mount Kepler. The eastern and highest peak, called East Peak, later was named Mount Copernicus.

By mid-November Fraser could report to the "Captain" that he was "hammering away at the south peak." "It looks like the bottom of a ship turned upside down. I will have an Island of it next week as I am cutting through just where Burnham's observatory stood. My force to work on the hill is 13 able bodied men besides myself and I never worked so hard in my life. About 1500 lbs. of black powder and 100 pounds of giant powder will settle the south end. If you pray at night, please ask that I may have no rain here till the south end of Mt. Hamilton is level to your grade. If you can keep it up till the first week in December, I will be all right." [37] They worked until December 21; then the weather became so stormy that Fraser reduced his force to four men and left for San Francisco to attend to other Trust matters. He returned in April 1881, working with from fourteen to twenty-four men. The rock was hard to drill. Blasting with five kegs of black powder, they broke out 900 tons of rock. In supervising the blasting, Fraser was as "careful as it was possible to be." They had no accident of any kind. "My men," he wrote, "works very hard as I stay [with] them and see that they earn their money. The men average apiece about 30 tons per day I took the scales up to the hill and weighed the loads occasionally and counted the loads. One large load I found to weigh 768 pounds." [38]

In this work Fraser after 1880 relied on foreman John McDonald, who was to become a valuable figure in the building of the observatory. A skilled machinist and a good all-around mechanic, he helped to level the top of Mount Hamilton, install the water system and power plant, and build the reservoirs and houses for residences, shops, and barns. In 1885 McDonald brought a bride to the mountain. Their daughter, Mary, was the first child born to a member of the Lick staff. Long afterward this daughter was married on Mount Hamilton to the astronomer Ralph E. Wilson.

On April 24, 1881, Floyd reported to Newcomb on their progress. He wrote, "I have not bothered you with a letter for so long that I begin to feel lonesome" and added, "I have just started work on Mountain again full blast and everything is being pushed with energy." He went on, "You would be surprised to see Mt. Hamilton now. The squaring down of the top makes a most beautiful and spacious deck, opening the magnificent view of the whole region by a simple turn around on one's heel. . . . Our water arrangements work splendidly and you can now have as many baths a day on Mt. Hamilton as at the Palace Hotel. The spring now runs about 5000 gallons per day as against 800 gallons per day towards the last of the dry season."[39]

Floyd also wrote exuberantly about the superb climate: "The advantages of atmosphere will not only make any given telescope 25% more powerful than any now situated elsewhere, but will make a given observer, or astronomer, I think, 50% more powerful than he could be at any observatory existing. I have lately returned from a most delightful trip to the summit." In the same vein he told Holden of the enchanting climate, the floods of water for bathing: "umbrella useless, and a superb atmosphere that fills in all the necessaries for the most aspiring summer of physical and heavenly delights."

In June the Lick trustees – Mastick, Plum, and Sherman – returned with Mathews to inspect the work. All were duly impressed. Fraser entertained them to the best of his ability as they explored the mountain from East Peak to Isabel Creek and Long Branch, taking time out to catch a dozen large trout and twice as many small ones. Afterward they complained about the roughness of the trip. In the evening, before falling into bed, they observed a lunar eclipse. The next morning, to Fraser's dismay, trouble erupted in the kitchen over the weak coffee the cook and his wife served to his men. He exploded, "I find that my cooks is tartars and must make a change and I see that we must come to Chinese. It is next to impossible to please working men as Cooks will not take the pains to cook for them that they ought to take. It is very bad for me to have a quarl at this time with the trustees here but I will stop the eternal grumbling if I have to discharge every man on the place as long as the trustees keep me here. I will have law and order . . . or know why."[40]

While these and other projects were being launched, Floyd felt he should be on the mountain. Fraser had set up a telephone line to San Jose, opening communication with the outside world, but this was not the same as being on the spot. In the late summer of 1881 Floyd moved to the "hill" to stay into the fall. Fraser was overjoyed. "Captain was just the man I wanted to consult in our work as it proceeded. He sacrificed a great deal of himself by staying as he neglected his own business for that of the Trust."[41] Inevitably Floyd felt the strain.

Another hard, if less hazardous, problem was the making of bricks. Luckily, two and a half miles from the summit on the Mount Hamilton road, Fraser found clay beds that might be usable. He made a contract with T. W. Peterson to burn a kiln of 160,000 bricks to test the clay. The test was successful, "an enormous piece of good fortune." A year later the trustees let a contract to deliver a million additional bricks, which arrived at the rate of 10,000 a day. In August 1881, Fraser began to lay bricks for the foundation and building for the 12-inch refractor that Floyd had bought for the Lick trustees from Henry Draper, one of the finest ever made by Alvan Clark. On August 23 he laid the first brick, and under it he placed a dime and a nickel.[42] When it was finished he crowed, "I consider the Pier one of the firmest ever built to support a 12-inch equatorial."

As superintendent, Fraser proved a stern master. Yet, if he drove his men hard, he drove himself harder. When he decided that an average of 1,200 bricks a day per man was not enough, he determined to find out from the foreman if there were any loafers. "If there is any, I will ship them quick as loafers will not do up here."[43] If a man claimed illness and Fraser suspected drinking, that man would be shipped off the mountain. In his "log" Fraser wrote, "I will soon get clear of whiskey men as I hate a whiskey drinker." (Fraser was a teetotaler. Floyd was anything but, yet this never seemed to affect their friendship.) Somehow Fraser found enough able men to carry on the work, and in record time the work was finished.

At times, of course, with the large number of men on the job, there was danger of serious accident. Dr. J. C. Shorb, a former army doctor and a friend of Floyd's, had sent a medicine chest – an ample one, as it had to be, because of the "great and inconvenient distance from the nearest druggist."[44] Once the carpenter was "taken by the colic very bad." Fraser administered thirty-five drops of laudanum and a teaspoonful of hartshorn, repeating the dosage until the cramps subsided. With a large dose of castor oil and peppermint, the man was out of danger before eleven that night. Without the medicine chest Fraser was sure the man would have died.

In a more serious case, Robert Cross, the "boss bricklayer," fell from the staging in the library, broke two ribs, and "hurt himself inside."[45] Fraser, afraid Cross might not pull through, stayed up all night with him. Then he called for a doctor. For a month Fraser was forced to serve as doctor and nurse before Cross, still weak, could leave for San Francisco. "Thank God he is gone. He was a terrible bore and trouble to me, and I hope never to have the like from a almost stranger again."[46] A few days later Fraser's wife dreamed Cross was dead. Fraser, on his way to San Francisco on business, stopped at Smith Creek and learned that Cross had died at noon of heart disease. The grateful bricklayer left him $400.

Such were the hazards on this remote mountaintop. Still there were many compensations, as Fraser and Floyd reveled in the natural beauty and the glorious mountain nights. One day, after a night of observing, Floyd exclaimed, "The Heavens last night were truly magnificent." In September 1882, when a great comet appeared, Fraser, observing it with delight, called it the grandest sight since Donati's comet of 1858, which he had seen over Cape Breton.[47] Even the many days and nights that brought weather of a different sort had their allure. On October 10, 1880, for example, Fraser's "log" runs, "Blowing hard, blowing a 2 reef breeze making things squeak like aboard ship."[48] There were days when it seemed they might be blown off the mountain. Yet the work went on regardless of the weather. They calculated the dimensions of buildings, the making and cost of bricks, and the amount of Vermont marble needed for the corridor in the main building. When, after spending some weeks on the mountain, Floyd left for San Francisco, Fraser wrote in his "log," "I will miss him very much indeed." Yet again, when Floyd asked him to come to San Francisco, the conscientious superintendent felt forced to answer, "I would like very much to see you, but, Captain, the work being done is of such a nature that I cannot leave except by death and positive orders from you to report forthwith. Of course if you send such orders I will be there without fail if alive, you can bet on that. . . . I am here to do what I am told, and, dear Captain, you will always find me as I have always been, to attend to your instructions as near a mortal can."[49]

The more they worked together, the more Floyd valued Fraser's work and his devoted friendship. He knew that once Tom understood every aspect of a problem and had his head "fixed straight on," nothing could make him deviate from "the greatest care and exactness–absolutely loving care and thoroughness."[50]

The problems Fraser faced were of all kinds. At times he was baffled. One day he struggled for hours to fill a mercury barometer. When he finally succeeded, he exulted, "I feel like Wellington, when he cleaned out Napoleon at the Waterloo affair and saved all Europe from a tyrants sway, only I had no Blücher to assist. Thank God, I done the job."

Floyd realized that the brunt of the physical labor fell on his friend. Once when a month passed without a word from Fraser, he wrote in alarm. Reassured by the news that Fraser was "sure enough alive and well," Floyd hurried to the "hill," stopping only once to rest on his way up the Mount Hamilton road. He had hoped to start sooner, but his good friends, Barty Shorb, Tom Madden, and Judge Ross were at Kono Tayee, and he felt he could not easily leave so much company or Cora, who wanted him there to host the party.

Thus Floyd was constantly torn between his social, banking, shipping, and other business affairs and his obligation to the Lick Trust. The conflict was one he could never fully resolve. When he promised James Lick to do his best as president of the Trust he had little idea what it would mean to him and to his family and friends. One day he confided to Howard Grubb that he must put off the private telescope he had planned. "In truth I fear I shall not want to look at any thing of the kind should I ever pull through the worries of the L. O." [51]

How often he longed to be at Kono Tayee when he had to be in San Francisco on Lick Trust affairs and those of Cora's estate. Once he wrote in frustration to Jack Fraser, "By Jove! This kind of day makes me wish that I was up there with you without any cares to worry me. Us only! and the fellow to cook for us. What lovely tinkering we could do! and such a fire and grog to toast our toes and warm our yarns with after eight bells at night. I hanker with watering mouth to think of it! but it's no use my boy can't come! I am, and have been for ten days, wrastling in wretchedness with one of the stupendous influenzas and the City's stupendous taxes. I must be here some of the time to stand like a man my share of the troubles of the body and of the mind."[52]

In 1882 he bought the Quercus Ranch across Clear Lake from Kono Tayee. Now he hated to be away from home more than ever. He told Tom Fraser of his joy. "The place here looks more beautyful than I have ever seen it and Jack [Fraser] has everything in good shape." The *City of Lakeport,* the steamer they had built and for which Floyd had received the mail contract between Lakeport and East Lake, hove in sight, and he ended his letter abruptly:

> The bloody steamer heaves in sight!
> And now her screeching whistle blows,
> I must hasten this with all my might.
> To send it by the mail that goes![53]

In the winter, when the Floyd family moved to San Francisco and 415 First Street, the trip to Mount Hamilton was easier. But he found the business and social demands greater, the conflicts more intense. He once wrote to Holden, "My private affairs load me with many distractions that only permit me to give my mind at intervals to that which requires from any man his whole undivided attention."

Thus, while Tom Fraser was and could be completely devoted to the project, Floyd, equally concerned, could not be so single-minded. He fully realized how lucky it was that Fraser was superintendent through all the years of building. At times he even envied him!

One of the hardest parts of Floyd's job as president of the Lick Trust, and the part he abhorred, was the endless correspondence. All his life he had hated to write letters. Now there was no other course. He told his brother-in-law, "When there is the slightest show to be criminally neglectful in the way of writing letters I cannot resist the temptation. If there's a hell for that kind of delinquent, I am bound there, beyond redemption." He looked forward to the day when telephone and telegraph would do away with nearly all letter writing. "They must of necessity be brief and to the point which will save such a lot of suffering – and such a lot of infernal nonsense." [54] Still his letters reflect in a remarkable way his strong personality, his boisterous humor, his love of life. Often, when he was in a hurry, he wrote in a scrawl, but, like his father, he wrote equally often in a beautiful copperplate hand.

6

Ladder to the sky

On July 5, 1881, Floyd sent Fraser a "despatch" he had received from "Warner & Swasey–Chicago."[1] He supposed this to be a machinery firm, wanting to bid on their dome work, "as it appears to be a consequence of my letter of June 22 to Professor Holden wherein I described what we are doing and proposed to do about dome and mentioned that I had sent plans to the Clarks to see what kind of bids we could get in the East for all the iron work or for just the rolling gear."[2] He added, "If the Chicago people make low enough bids I think they would probably be the most pushing of Eastern firms and we would have the advantage of having them stirred up by both Holden and Burnham." In their "despatch," Warner and Swasey offered to machine the running gear and rack for the revolving 12-inch dome and to finish it by September 1, in time for the transit of Mercury on November 7, 1881. Soon after this a letter arrived from Holden. He called them just the men for the job. "Burnham and I both know them and they have sound ideas."[3] Burnham wrote equally enthusiastically: "I do not know of anyone else so likely to put it through on time, and do a first class job. I think you may count upon their doing anything they undertake."[4] By this time Mathews had already telegraphed official receipt of the order.[5] Floyd then asked Clark to forward the detailed drawings he had made with Fraser, based on their observation of domes here and abroad. The designs included the shutters, the dome covering, and a "Live Ring," on which the dome covering rolled and which, in turn, rolled on the circular wall and the loading-bearing cylindrical wall on which the "Live Ring" and dome rested.

All this followed a disappointing effort to have the work done by a San Francisco firm. Floyd had hoped to supervise the work there, making changes as the work progressed. But the estimates proved too high, and he was forced to go East with so little time that they had to resort to a gamble. Luckily the gamble paid off. Still he was unhappy. If he could have had his way, and if there had been time, he would have turned instead to Howard Grubb in Dublin. In fact, he was embarrassed that they were using many of Grubb's ideas in the wooden dome design, without his knowledge or permission. As he told Fraser, he felt "at a loss" as to how he would ever be able to show Grubb that it was "a liberty taken for the sake of science and not prompted by any unfair motives."[6]

When Grubb learned that the contract had been awarded to the firm headed by Worcester R. Warner and Ambrose Swasey and that he had not even been given the chance to bid on it, he was extremely upset. His firm lost both the contract and the designs, as one condition for submitting a bid was that the winner – in this case, Warner and Swasey – should have access to all the designs submitted. From the beginning he had been violently opposed to this system of competitive bidding, proposed by Simon Newcomb.[7] He thought it "lowered the stature of the work" and treated fine astronomical instruments "like agriculture implements or steam engines." He believed the design of the telescope should be "solely the privileged knowledge of the designer and not a debatable concept that the trustees or any other builder had a right to judge." He feared, with reason, that his original ideas on telescope design would, if offered in detail, "be vulnerable to assimilation into the design of a competitor or even incorporated into dissimilar plans after the contract was awarded." He felt strongly that the final instrument should be the expression of the ideas of a single designer. As it turned out, his fears were probably justified. But, above all, time and distance were against the Irishman.

When Floyd first heard from Holden about the Chicago firm he knew nothing about it. Later he learned more about the lives of the partners, Warner and Swasey. Both were New Englanders, both were born in 1846, Warner in the small town of Cummington, Massachusetts, Swasey in Exeter, New Hampshire. They soon became friends, and in 1869 they went to Hartford to work for Pratt and Whitney, manufacturers of firearms during the Civil War. Trained as machinists, in time they would be considered mechanical engineers.[8]

As a boy Warner had become interested in telescope making and dreamed of the day when he could manufacture telescopes. From 1869 to 1880, with Swasey, he worked his way up in the company, developing managerial and contracting abilities, until they had built capital to found their own firm. Whereas Swasey would be recognized for his gear-cutting ability, Warner would become known for his talent as a salesman. In 1876 at the great Philadelphia Exposition, Warner was put in charge of the Pratt and Whitney machine tool exhibit. One day, after setting up the exhibit, he decided to take time off to go to Washington to see the recently completed 26-inch telescope at the Naval Observatory, then the largest in the world. He boarded the midnight train and arrived at the observatory to find that visitors were admitted only on Saturdays. He pleaded with the janitor, who had no authority to admit him, to let him see the telescope. Just then a young man appeared at the door and asked if he could help. Warner told him of his interest in astronomy and asked if he could see the "great" telescope. The young man was the assistant astronomer, Edward S. Holden. Soon he was showing Warner around the observatory. So began an association that would have important repercussions on the Lick Observatory some years later.[9]

Warner returned to Philadelphia. Soon afterward he asked Holden if he could use the instruments the navy was displaying there. These instruments, used at the 1874 transit of Venus, included an equatorially mounted, portable 5-inch Clark refractor. Warner then asked to borrow a set of keys to the observatory where the telescope was mounted. Soon the keys arrived. All through that summer Warner came and went in the evenings, studying the 5-inch refractor as well as a 7-inch equatorially mounted refractor made by the Washington firm of Fauth and Company. Unlike most contemporary telescopes, the 7-inch had eye-end controls by which the observer could control and guide the instrument. Warner's studies during that summer must have had a tremendous effect on his thinking and on his future ideas of telescope design.

In May 1880 Warner and Swasey had moved to Chicago, to a shop on the west bank of the Chicago River. Here, in addition to their more traditional lathes and milling machines, they began to advertise "fine work on astronomical Instruments." Early in 1881 Warner learned that Holden had become director of the Washburn Observatory in Madison. Soon he asked the astronomer to examine the 9½-inch telescope that had just been completed. Holden, duly impressed, persuaded Beloit College in Wisconsin to buy the instrument. The partners must have been elated. This gave them the vital entry they wanted into the telescope field. It was also a step toward the much greater dream of building the mounting and dome for Lick's giant 36-inch telescope on Mount Hamilton. At this point that was a distant dream. The first important step toward that goal was the winning of the contract on the dome for the 12-inch refractor.[10]

When the 12-inch telescope arrived on Mount Hamilton in October, 1881, Fraser had nearly finished the brick work for the dome foundation and the transit house. As Fraser and Floyd rushed to get ready for the transit of Mercury, time was of the essence. They felt the pressure, and one of their rare disagreements resulted. In general Fraser tried to follow the Captain's instructions as closely as he could. When Floyd was on the mountain, he relied on his advice. When he was not, Fraser occasionally took on responsibility Floyd considered unwarranted. Once, in connection with the rolling gear, Fraser, in a hurry as usual, telegraphed Warner and Swasey about some problem. The quick-tempered Floyd exploded. "You should on no account have sent that telegraph except through me. . . . Never mind how right the object or how important might be time, the confusion resulting from important orders from more than one on the same subject defeats the very object you sought in gaining time. Your despatch makes a condition of things where it is now uncertain that any despatch could be sent to Warner and Swasey that would be effective – because they would be uncertain about the authority. All despatches to me on Lick Trust business will be sent from Lakeport immediately to Kono Tayee by Captain Bundy or other boat man." [11] This was certainly a roundabout route; Fraser, who liked to get things done "yesterday," often thought it quicker and easier to act on his own.

Floyd ended his sermon by saying he had sent a telegram to Warner and Swasey telling them he would trust their judgment. "It is too late to give any more instructions now – it would only botch things, destroy their ambition to please, and open wide the door for extra charges. If let alone their machine will work well. I'll bet as well if not better than Grubb's." [12]

At the end of October Floyd learned that the rolling gear was on its way to California and that Warner was coming out on his own hook; that is, the trustees had told him they could not pay his expenses but would be glad to entertain him "whilst here." Holden called him "one of the most talented mechanics in our country – is an astronomer, is very ambitious" and "with all is an exceedingly pleasant gentleman." [13] Floyd asked William Sherman at the Lick office to "have him taken care of in the very nice way that I know you so well understand, and ship him off to the Observatory in good shipshape style." [14] He added significantly, "I think Warner comes out to make a study with a view to ultimately bidding for the construction of our great Dome and mounting – the which may result to our advantage." He concluded, "We are doing splendidly with our work and will be ready for the transit in good shape." He hoped all the trustees, including Sherman, would be there for the event.

In November Warner arrived on the mountain to install the gears for the 12-inch dome. Eager to press his firm's claims through closer contact with Floyd, he confided to Swasey, "By seeing so much of the Captain I can learn his ideas and how to deal with him."[15] Whenever possible he tried to impress Floyd with the knowledge he had obtained on his trips to Europe and his study of European and American observatories, including his visit to Grubb in 1878, where he must have gained valuable information on the important problem of telescope engineering. Already he felt confident his firm could "get" both the mounting and dome of the 36-inch. A skillful salesman, and a bit of a con man, he had realized early the advantage of cultivating not only the president, but also the other Lick trustees.[16]

Warner explained that in their dome design, he and Swasey had used a method of applying power to the revolution that they had used on spinning frames in their textile machinery work, where they had to balance two forces so as not to break the yarn. For the dome they proposed that, in a similar way, "the power he applied on two diametrically opposite points at the base of the dome frame." After the revolving gear for the 12-inch dome proved successful, Warner informed Fraser with great assurance, "I think we could design a running gear for a dome of any diameter and make it so that it would run perfectly."[17] Soon, however, Fraser became suspicious of Warner's extreme self-confidence, his overweening salesmanship. In his "log" he noted that the plans sent to Warner by Clark were almost the same as those followed in the final design, worked out by Fraser and Floyd. "Except for the revolving machinery the sketch for latteral wheels is all our own idea."[18]

The partners' knowledge of machine tools and instrument making was great; yet they surely realized that it was a tremendous leap from the 12-inch dome to the giant dome for the 36-inch, or from the mounting for a 6- or $9\frac{1}{2}$-inch telescope to the monster for the 36-inch. At times, they too must have had their doubts. There was no great difficulty with the small telescopes, "but every increase in size adds a new problem of how to make it work conveniently, smoothly and successfully and follow a star without vibrating."

Meanwhile, soon after the arrival of the 12-inch telescope on Mount Hamilton, Floyd learned that Kalakaua, the flamboyant, dark-skinned king of the Sandwich Islands, whom Floyd had known there long before, had arrived in San Francisco. He had visited the Floyds soon after Harry was born in 1873; he called her "Kalikoolani." Staples, acting president of the Society of California Pioneers at the time, had welcomed the king to San Francisco. Knowing of his fondness for champagne, oysters, and beautiful women, he "got up" a dinner in his honor. Afterward Kalakaua asked to meet James Lick.[19]

When Staples told Lick that the king wanted to visit him, the old man said he did not want to see any "niggers." But the following day he changed his mind. Staples noted that he had evidently been thinking of the king's coming to see him, and the idea gratified his vanity. Staples told him to have his wig fixed and his whiskers dyed, as the king would be there at one o'clock. He then hurried over to the Grand Hotel, where the king was staying with his retinue, and told Kalakaua that Lick would prize his autographed portrait. They arrived back at the Lick House, and Staples presented His Majesty. "Then that magnificent type of a man, stalwart fellow with black hair, splendid features and a bronzed complexion stood before Mr. Lick, and said that he had heard what Mr. Lick had done, and what he proposed to do for the state, and he thanked him on behalf of Humanity. All Lick could do was hold on to the King's hand with a tight grip. . . . Kalakaua then spoke to his secretary, who handed him the photograph. The King asked Lick if he would accept it, and Lick gripped hold of it. When the interview was over, the King backed out of the apartment in courtly style." As the party was leaving, Lick asked Staples if he was not going to give them some lunch. Staples told him that Schönewald was preparing a first-class repast for the party. "Well," said Lick, "give them the best there is, give them all there is in the house."

Now, as the king arrived on the mountain, arrayed in a fancy uniform, Fraser noted in his "log," "Capt. Floyd left to bring the King up today. One King, one load of sand." The Captain returned with the king and Cornell Judd at nine o'clock the same night.

Kalakaua arrived, as we have seen, at a crucial time, as the first important astronomical venture on Mount Hamilton was about to be launched. The 12-inch dome was not yet finished. Fraser therefore had to improvise by mounting the telescope temporarily on the pier in the open air. The king was almost the first to look through the telescope. He was entranced. They went to bed at 1:30 A.M. The next morning he was up at 9. After breakfast he again went on the "hill." He told Fraser he was delighted with what he saw and wanted a transit at his place. He left at noon with the Captain. Fraser concluded, "So much for a King."[20]

The following day the carpenters went back to work and life returned to normal. Ten days later Fraser mounted the 12-inch telescope in its new dome. It was revolved by an ingenious mechanism that Floyd and Fraser had devised. Quite simple but highly efficient, it moved silently and worked perfectly in all weather. Yet its cost was about a hundredth of that of the old-style ratchet and pinion. By means of an endless wire rope, attached loosely around the exterior and connected by pulleys to the outside, the labor of moving the dome was reduced to a minimum. This, said Floyd, was "all the business." It cost about $15 and could be easily turned by a child.

By this time Holden and Burnham had arrived on the mountain to observe the transit of Mercury, just in time to feel the force of the winds blowing over the mountain. In his "Log" Fraser noted, "Blew hard all night made things howell the day was too wild to work on hill." Instead they played poker.[21]

Burnham and Holden first met when Burnham visited the Naval Observatory in 1874. Luckily, on August 11 of that year, they had a night of "splendid seeing" when they observed the ring nebula in Lyra. On another night they examined the Trifid nebula. So began a friendship that would have an important impact on the history of the Lick Observatory.

Now, by November 7, 1881, everything was miraculously ready for the transit and the visit from Trustees Mastick, Plum, and George Schönewald, now manager of the magnificent new Del Monte Hotel in Monterey, and Secretary Mathews. Fraser showed them the moon through the 12-inch and said proudly, "Think it is fine and so says the Professors." When they finally went to bed they left Burnham to his "quiet and happy devotions in search of more worlds."

On "The Day we Celebrate!" as Mathews wrote of this, the first important astronomical event on the mountain, "Every glass was in readiness for the Transit of Mercury."[22] It was visible in the surveyor's transit, as well as with the 12-inch refractor. Said Mathews, "The observations were successful owing to the very fine day and to the excellent preparations made by Captain Floyd, the prest. of the Lick Trustees."[23]

After this, Holden's first visit to the mountain, Fraser called him, admiringly, a "boss man without a doubt."[24] When, in 1881, Holden succeeded J. C. Watson at the Washburn Observatory in Madison, Burnham noted that Holden had more "go aheaditiveness than any man in the astronomical business."[25] Floyd, too, was impressed. The Captain, then commodore of the San Francisco Yacht Club, sailed the astronomer across the Bay and afterward gave him several of his paintings, including one of the *Ariel,* his sailboat he raced on San Pablo Bay. They even talked of sailing together to the Sandwich Islands. Back in Madison, Holden, grateful to Floyd for this souvenir of a memorable time, hung the painting over his dining room sideboard. He wrote warmly, "It was a very charming and real sensation I had there and made an intimate and deep impression upon me, in which it is hard to separate the Astronomy the friends and the scenery."[26]

At the end of this busy year of 1881 Fraser left for San Francisco "bag and baggage." He had been on the "hill" for three months without leaving, taking star transits for time and making comet observations while also building the main observatory structure. In his "log" he wrote, "I am glad to get away. So ends the second chapter of the work on the L. O. Goodbye old Log. It is quite a relief to let you alone for a while for it is no fool of a job to write a log every night for months. Just try it, gentle reader, and you will find out for yourself. So goodbye again."[27]

A year later David P. Todd of the Naval Observatory arrived with photographer J. L. Lovell to observe the transit of Venus, the last such transit until the year 2004. On December 6, 1882, the sun beamed down on the tense group of observers anxiously awaiting the moment when Venus would appear as a black spot as it crossed the sun. Fraser, who helped Todd with his final adjustments, found him "bang crazy with his responsibility." [28] Everyone helped. Cora Floyd and Floretta Fraser recorded time in the transit room. Cora Matthews, Cora Floyd's niece from Louisiana, made drawings. Floyd recorded observations with the Clark 40-foot horizontal photoheliograph, set up in conjunction with the transit instrument.

When photography had first been used at the 1874 transit of Venus, observed from Siberia, China, and Japan to the Indian Ocean, it was hoped that, from these observations, the sun's distance could be more accurately determined. But the weather had been bad and the results were poor. Now, in 1882, the photographs were beautiful, but, as it turned out, the results proved inconclusive. Nonetheless, in his diary Todd noted, "Day as perfect as a June day in New England–sky perfectly cloudless. A day built only for the Gods and Mt. Hamilton astronomers. . . . We saw things as plain as was ever seen with any glass in the world." He added, "Our success gave the New Institution a grand send-off, greatly helping it, as Mr. Lick's Folly had been much ridiculed." [29]

Henry Mathews lined the observers along a fence and recorded the event with his omnipresent camera. Some were dressed in city clothes; others, in outfits more suitable to the mountaintop. Floyd was arrayed in a velvet suit; on his head he wore a helmet that made him look like a Viking. Among the observers was his friend William Welcker, whom he had known as the steamship agent in Oregon in his seafaring days, now a mathematics professor and regent of the University of California.

Floyd and Fraser had worked day and night to be ready for the transit. When Todd's report was published, Fraser felt he did not "tell enough of the trouble Floyd had taken, assisted by the Trustees to secure for science at this Observatory the Transit of Venus." [30] They had hoped that Newcomb, of the United States commission, might come out for the transit. Instead he went to the Cape of Good Hope, leaving them "out in the cold." Floyd, deeply disappointed, was forced to engage Todd, and to pay for the expedition out of his own pocket. Other commission members told Todd he would find nothing at Mount Hamilton with which to work. They even suggested sarcastically that he take a faucet "to put in the bowl for a tank in his Photo room." The photoheliograph had arrived from the Clarks with one end missing, and no focal length for the objective given. Clark, when asked for it, said he had forgotten, but he told Fraser how to find it. That way "failed utterly." Holden's way failed also. Fraser had set the piers according to the government instructions for the transit, only to find an error of two feet for their objective. Floyd "got a camera" and advised Fraser to get a piece of ground glass. Following Floyd's directions he adjusted the object glass and "Lo and behold there was the focal length 40 feet! – just as Todd made it afterwards, and not 38 feet as Clark and the government had directed."[31] As a result, when Floyd arrived on the mountain before the transit, the instrument was set and well adjusted, and he was able to take some pictures of the sun "in good shape" before the astronomers arrived.

A month later Fraser again closed his "Log" for the season. "Next summer will see more done if the Lord spares me to see it finished is all I ask but if not all right as I hope a Better man take my place in my absence on Jupiter as that is the Planet I have selected for my futter home. Good by Good by for this chapter."[32]

When Fraser returned to work on the "hill," one of his first visitors was James B. Francis of Lowell, Massachusetts, who, as chief trustee for the Uriah Atherton Boyden estate, had been left with more than $200,000 to build an observatory on some mountain site. Said Fraser, "He is about 68 years old and a fine man." Fraser brought him up in a "buggie" and showed him Jupiter and its satellites, as well as Uranus and Sirius. "He likes the place well."[33]

After this other visitors flocked to the mountain. Jules Janssen, whom Floyd had seen in Glasgow and Paris, arrived on his way back from the May 1883 eclipse at Caroline Island. With him was Léopold Trouvelot (whom Fraser found "no account"). Janssen, though "quite old," examined everything thoroughly and scribbled notes on the construction of the 12-inch dome and telescope.[34]

At times Fraser was overwhelmed by the character of those who came. When President W. T. Reid of the University of California arrived with a party of astronomers, including George Davidson, Fraser exclaimed, "My Lord, what a crowd on a mountain of intellect, worth to the world millions of ordinary men!"[35] Another day James Lick, nephew of the donor, turned up with a party of friends.

A contrast was F. M. Graham, the courageous bicyclist who made it to the end of the winding mountain road on the "first vehicle of this kind we have ever seen here."[36] Fraser welcomed him warmly, as he did other visitors, ranging from Albert Bierstadt, the romantic artist, to George Wharton James, the writer who described the observatory in his article "How We Climb to the Stars."[37] James was particularly struck by the light and airy appearance of the 12-inch dome. He contrasted it with the heavy, clumsy look of most domes he had seen: "Its beautiful curves are not marred by massive iron work or the ugly bracings generally used." He admired its framework of steamed bent oak, covered by copper sheeting, plated with tin inside and nickel outside. Above all he praised Floyd and Fraser, those "men of inventive genius" who, though not experienced dome builders, had shown wonderful skill in leaving the beaten track and blazing new ones.

Floyd, too, was pleased with the "very clean, solid and cheerful appearance" of the "very handsome" dome. He called it a "bijou" of its kind.[38] Next to the dome he had furnished a cozy den, with a little stove. There he often worked with Fraser far into the night, hammering out the details of plans for the great 75-foot dome and the three-story astronomers' brick dwelling, intended for the director and the astronomers in residence, as well as plans for the library and the house for the meridian circle (that fundamental instrument needed to determine the exact position of heavenly objects as they cross the meridian). As Floyd told Newcomb, he would like to be at the observatory all winter "if it were not for my family and my business which makes it impossible." He added, "It seems a pity to let so many magnificent nights as we have been having during the Winter go by without an eye to behold the beautiful things that must have passed through the field of our 12-in." [39]

When the meridian circle house, with its roof opened by hinged shutters – largely Fraser's design – was finished, he wrote "The meridian circle shutter works fine and I think it is a great success and I must give myself the credit of it as it was done against some Professors' wishes, but Captain Floyd stood by me and we have got the best in the world in the shape of a meridian circle all through, and it will be hard to beat it anywhere." (One of the professors to whom Fraser referred was Holden.) It was a handsome building with double walls of iron and wood to provide an equable temperature and solid masonry foundations on which the meridian circle, made by the Repsolds, would be mounted. He told the Captain, "I can show a reasonable man like yourself, that Holden jumps at conclusions." [40] Holden, he said, could afford to make mistakes on account of his national reputation, "when we can't." In time, Holden's errors would become their errors, and all the blame would fall first on Floyd, then on Fraser. When the job was finished, Floyd was delighted. "The work looks fine indeed!" he exclaimed. When his roof later weathered some severe storms without leaking, Fraser was overjoyed. He claimed that this was "thanks to no architects having a hand in it."

Looking back four years later, Fraser noted with a tinge of arrogance, fed by disillusion with Holden, "At the time the Meridian Circle House was built Capt. Floyd and myself had made such a close study of what was required for a great Observatory, that there was little need of advice from Astronomers; of course we were always ready to receive friendly counsel and Prof. Holden being our supposed friend suggested many things that was thoroughly considered as we supposed they were kindly meant

Thomas E. Fraser with the meridan circle on Mount Hamilton. (Reproduced by kind permission of the Mary Lea Shane Archives of the Lick Observatory.)

to assist us in our great work, but there was few things we did not work out ourselves from the views of various Astronomers, not from one. The most assistance we got in our work was from a translation of the works of the Pulkowa Observatory made in Russia." When Holden later claimed responsibility for much of the work, Fraser wrote angrily, "We don't intend at this date to allow any person to say they furnished completed plans for the Observatory; the plans for the Observatory were furnished by the best brains of men of the whole world, but these ideas were crude. We put them together – the Lick Trustees, Capt. Floyd and myself saw that they were well weighed and digested before they were carried to a final conclusion."[41]

As Edward Pershey, the Warner and Swasey historian, points out, "Floyd's ability to assess the instrumentational needs of astronomers and grasp the concept of a group of specialized telescopes for the Lick Observatory was an impressive and significant accomplishment in his role as President of the Trustees. His sensitivity to and acquired knowledge of this field were important in drawing up the design and construction contracts for the large telescopes in the early and mid-eighties." [42]

So the work went rapidly on and Fraser was "as busy as a rattler." During the summer and fall of 1882, most of the main building between the 12-inch and the future 36-inch dome was laid and roofed. Toward the end of the busy June of 1883, Fraser prepared to lay the cornerstone under the southwest corner of the main entrance. On July 8, he placed copies of the San Francisco and San Jose newspapers, some silver coins of the United States, a photograph of James Lick, and Lick's spectacles in a copper box. All the workmen and other mountain inhabitants gathered to watch him put the box in the cornerstone. He must have been disappointed that neither any of the trustees nor the Captain was there. The "humble Superintendent" gave a short speech. "The time will come," he said dramatically, "when this observatory will allow none to outrank her in the advancement of science. If our large objective proves a success, which we have every reason to believe it will, then this place will stand unsurpassed probably for years and years to come, as having a more powerful instrument with which to inspect the billions of suns that are in the distant firmament, beyond these stars that we see shine out so brightly at night with the naked eye. May not each of them be a distant sun with worlds revolving around it like our own system? Our sun has worlds swinging around it controlled by the law of attraction. Neptune, the farthest away of our system that is at present known, is at times distant 1,800,000,000 miles from us. Just think of it: Is every star we see a sun like ours with worlds revolving around it?" [43]

He went on: "This work is the greatest undertaking in the realm of science that has ever been thought of. The empire of Russia sent out the most eminent scientist in the kingdom, Prof. Struve of Pulkowa, to investigate and contract with the most skilled workmen that could be found in the world for the manufacture of the largest refracting object glass that, in his judgment, could be made to work satisfactorily. After travelling over Europe, he came to this country, and in a small workshop in Cambridge, Massachusetts he made a contract with Alvan Clark & Sons to construct what he, Struve, thought would be the largest refractor in the world."[44]

But trustee Plum, in that summer of 1883, was worried about Fraser. He was afraid he was not taking proper care of himself. "You are too anxious to accomplish a great deal in too short a time and I fear you will give out before you are through. . . . Don't put off too long your needed recreation. You will not do any good by overdoing nature and then being broken up the rest of your life." He added, "I hope Mrs. Fraser will go for you and assert her power before it is too late." But Fraser was too busy to heed Plum's advice.

So gradually the mountaintop took on the look of a small village. From a distance it resembled an ancient castle. Yet the heart of the telescope was still missing; without it there could be no telescope. With comparative ease the Feils, in Paris, had cast the flint disk, a flawless mass weighing 700 kilograms, over 38 inches across. In April 1882 they had shipped it to Alvan Clark. Since then, months had dragged by without success on the crown component. Floyd was becoming increasingly frantic. Newcomb considered the chief cause of failure so simple that it should not offer any trouble once it was detected.[45] When the founder succeeded in casting the 36-inch lump of glass, weighing several hundred pounds, the clay pot in which it was contained had to be broken off, together with the outer part of the glass itself that was impregnated with clay and other impurities. This, a long, tedious process, was done with a wire, working in sand and water. It could take weeks, even months. The glass was then heated near the melting point, the lump was pressed into a disk, shaped like a grindstone. Here the trouble came. As Feil heated the huge lump it flew to pieces. Again and again he tried, taking more time for the heating. Again and again he failed.

Plagued by a press clamoring for results, Floyd wondered if the telescope would ever be finished. The months dragged into years and the agony continued. In the summer of 1883 Newcomb reported from Paris that Feil had the second crown glass nearly ready when in February it followed its predecessor. When the glass was being reheated after the outside had been cut away, it broke. Veins had to be cut out, the block reheated and reannealed. This would take more months.

When this disastrous news reached Kono Tayee, Floyd decided to forget the observatory entirely. He took off on a hunting expedition in the mountains with Cora. Later he apologized for the delay in his correspondence, caused by her enthusiastic mania to kill another buck with his telescopic rifle, "Hungry Tigress." For three weeks they did nothing but hunt deer by day. Cora walked every man in the party nearly to death.[46] At night they slept on the bare ground. They reached Kono Tayee in time for Christmas, and the ten-year-old Harry entertained family and friends with amateur theatricals. Afterward they headed for San Francisco for the winter. They arrived there after a "romantic drive" with the stagecoach driver "Joe Johns" and his six stalwart horses through mud and rain, while Harry, sitting on the box in one of her father's overcoats, a sou'wester, and two of his raincoats, screamed with delight.[47]

After this interlude Floyd learned that the situation in Paris was even worse than he had feared. It was now obvious that human as well as technical factors accounted for the Feils' continuing failures. Newcomb found out that the younger Feil, who had taken over the plant when his father retired, was an alcoholic. In December 1884, George Clark turned up unexpectedly at the Feil factory. He discovered young Feil well filled with rum beside a 600-pound lump of glass that *might* prove successful.[48] Eventually the heartbroken elder Feil had to return to manage the plant. He wrote sadly that he saw his son no more. For more than a month his son had not been in the factory. He considered him lost "by the bad resorts and by his insubordination." Still the elder Feil hoped to avoid total disaster and produce the promised disk.[49]

Floyd, although sorry for the father, was disgusted: "The Observatory has been turned into an 'elephant' by Feil's failure."

He feared it would be "a whole menagerie before the big telescope was finished." When Holden now urged that work begin on the foundation for the great dome, Floyd said he did not feel free to go ahead with such plans until the heart of the telescope was ensured. He wished he could make it useful in the ways Holden suggested, but "Alas! It is not my Observatory." [50]

Months dragged by. In the spring of 1884 Floyd told Newcomb, who was in Europe, that Feil had hoped to ship the crown glass during the first days of the previous September. Every day since then he had been expecting a telegram announcing its safe arrival. By this time Newcomb knew, as did Floyd, that they could not believe a word Feil said until the thing was actually done and aboard ship. Three months later the crown glass still had not arrived. Instead Clark sent copies of letters from a Paris friend who had visited the Feil foundry. For Floyd these, to some extent, lifted "the veil which has for so long covered our crown disk in mystery, and caused me profound anxiety." [51]

That winter one of the worst storms in history struck the region. At Kono Tayee it hit at 4:00 A.M. on December 26. For ten days it raged. On the tenth day it was "blowing a whooper from N.W. with the lake as rough as blazes." It rained 8.06 inches. The lake, up nearly two feet, was still rising. The roads were washed out, and the stages could not get through. Floyd reported to Fraser, "The Ukiah stage drowned a horse. Joe Johns was nearly swept away with the whole business in the little creek near Middletown, and the Cloverdale has taken two days to get through. All freight teams have stopped. I don't know when the roads will let us come down." [52] He was worried, too, about conditions on Mount Hamilton. His concern was finally relieved by a letter from Fraser reporting the abatement of the storm after eleven days hammering at his work "with all the violence the elements could produce." There had been $32\frac{1}{2}$ inches of rain in all. With three men on the mountain, the faithful Fraser had watched day and night. "Thank the Lord," he exclaimed. "It is over and I can have a rest." He concluded simply, "We saved the place by taking care of it." [53] Floyd knew what that meant!

Nonetheless, Floyd tried to take the storm philosophically. As long as nothing intended to be permanent had been blown or washed away, he felt the storm was worth more to them than all the plaster. Clearly they needed to redouble their preparations for any calamity. He called on his years of experience with the weather at sea and his more recent experience on Mount Hamilton and at Kono Tayee, where, since 1873, he had kept a daily record of the changes in wind and temperature and barometric pressure. He tried to think of every conceivable way of protecting the observatory and its inhabitants from rain, wind, and snow. Among other things he suggested a porch with double doors at the north entrance and a vestibule where the mud and snow from boots could be left: "a necessity in buildings for delicate instruments and especially so on the summit of a mountain reaching up into the snow and rain clouds of the prevailing storms."[54]

All through those stormy days Floyd had been doing lots of pen work. In contrast to the weather he "had been most damnably dry . . . as I have flown the blue pennant from my main ever since I left a vermillion orgie in S.F., and as this is a reform administration I don't propose to haul it down (at least in home waters) until the next Presidential election." [55] News of the alcoholic Feil in Paris and the resulting problems had driven Floyd himself to drink, and he was now atoning for his sins.

At this point, it seemed, all the elements, both human and celestial, were against Mr. Lick's "magnificent enterprise." Floyd and Fraser could overcome the problems of weather, hoping the difficulties over the crown glass would soon be resolved. But, as they fought to finish the observatory in the face of unexpected problems and tremendous odds, the hardest thing to accept was the growing criticism from outsiders, who felt the work was moving too slowly. This criticism, often initiated by people who, for one reason or another, were disgruntled, was avidly gobbled up by a press looking for sensation. Realizing the source, Floyd tried to ignore it, but could not. At one point, when the attacks reached a crescendo, often becoming personal, Holden told him that any gentleman could afford to laugh at such things. "For yourself personally you know every step has been guarded, and your military training has been of immense daily service to you in keeping you in the best possible trim to repel attacks of that kind."[56]

Vociferous opposition continued to come from Davidson at the California Academy of Sciences and from the Society of California Pioneers, Lick's residuary legatees, who were clamoring for their inheritance. On October 2, 1883, the Bradford Committee of the Pioneers blasted the Lick trustees. The *San Francisco Call* reported its actions. The Pioneers argued that, if the estate had been settled within two or three years it would have saved the residuary beneficiaries nearly half a million dollars. But, instead, they claimed, a large part of the income had been swallowed up in repairs, salaries, and "reckless mismanagement." They demanded to know when the trustees were going to sell the property and settle the estate. They wanted their money right away. They would not need it when they were "under the sod"! In fact, Mr. A. C. Bradford wanted *all* the Lick trustees removed from office. He urged the Academy of Sciences to concur.

Finally Lick trustee William Sherman, a Pioneer himself, was allowed to speak. He pointed out that there were at least fifty errors in the report, which had been prepared by a man who had sought a job as legal counsel for the Lick trustees and been rebuffed. Sherman presented the facts as recorded in the financial report of the Lick Trust from December 1, 1876, to October 1, 1883. It showed a profit of over $450,000. After the meeting Floyd wrote gratefully, "If there are enough just men among them as I hope, you will change the evil opinion the malice designs, into honor for the Lick Trustees!"[57]

Support for Floyd and the Lick trustees came, too, in the November 9, 1883, issue of *Science,* probably from the pen of Edward Holden. Over the years since the formation of the Lick Trust the Lick properties had increased greatly in value. If those properties had been sold when litigation ended, the trustees would have incurred a great loss. There would have been no money to divide between the Pioneers and the Academy. Even some specific bequests would have remained unfulfilled. The trustees, the author noted, had managed the Lick estate as carefully as possible and had sold only when it was to their advantage. From December 1, 1876, to October 1, 1883, the aggregate net profits had been $453,458 – over $60,000 a year. The author concluded that the residuary legatees now had $192,000 to divide. "Not long since they had nothing."

At this point Floyd finally decided to go to court to ask (contrary to the terms of Lick's will) for a partial distribution of the estate before the observatory was completed, "to eliminate a pressure that might force its construction to injurious haste." "We hope," he said, "to complete the Observatory safely as soon as possible."

In the face of all these difficulties, Floyd wrote to Todd, "I am really run to death with correspondence and business matters and I do most heartily wish that I was out of this Lick Trust business. I am worn out and disgusted with it and besides, it is a great expense to me, for which I gather only abuse and trouble."[58]

His problems persisted. Two years later he was deep in negotiation over the Lick House and Santa Catalina Island. A chief asset of the Lick estate was the Lick House. In October 1885, the trustees were offered $1,050,000 in cash for the property, but they had set $1,150,000 as their lowest figure. One day Floyd wrote frantically to Holden. On Santa Catalina Island problems had been raised about leasing a mine. On October 10 Fraser sailed for Catalina after Trustee Mastick warned there might be trouble with the "gold-fiend," H. A. Frey, growing out of the contract made with him in reference to his alleged mine discovery on Santa Catalina Island. Mastick was satisfied that there was no mineral there, but felt that the trustees might be required to prove it. The trustees decided to send Fraser down to investigate.

7

Success and conflict

Early in 1885 Floyd had planned to go to Washington. Instead he sent Fraser who, after a lively train trip, arrived there the day before President Grover Cleveland's inauguration. Amazed by the huge crowds and the excitement, Fraser called on Newcomb in the Nautical Almanac Office at the Navy Department, in the State, War, and Navy building. From that office they watched the spectacular fireworks in the evening. Later, Fraser, an ardent Democrat, met with Cleveland at the White House. The president asked him "all about the Observatory and Mr. Lick," and Fraser told him about Floyd.[1] Cleveland said he would go to California to look through the telescope.

Later Fraser met Holden and visited the firm of Fauth and Company, makers of the 6-inch Lick transit. They went on to Charlottesville, Virginia, where Warner and Swasey, improving on the 25-foot dome for the 12-inch refractor, designed by Floyd and Fraser, had won the contract for the 45-foot dome at the Leander McCormick Observatory of the University of Virginia. The telescope had been mounted the previous year. Fraser and Holden were impressed. Fraser was also impressed by Holden: "He is a Delightful Traveling Companion and he is head and ears over all the astronomers I have met." Fraser found the observatory director, Ormond Stone, "very nice–but will never be above the grade of Todd, for instance," He found the work there in keeping with most things in the South, "long behind the age," except the dome and the telescope.[2] As Fraser continued his travels, he visited most of the leading observatories in the East. Everywhere, he said, he tried to see the good but also the bad points, in order to discuss them with Floyd afterward.

In Cambridgeport he found the "old gent," Alvan Clark, more feeble but still working with two fingers stitched up from rubbing glasses.[3] In the yard behind Clark's shop stood the great 30-inch refractor for the Pulkowa Observatory. Clark said he had made every test on it himself but could not do the same for the Lick telescope. He told Fraser that Feil was working on another disk and hoped for the best. He had found a wealthy man named Mantois and, with his help, expected to carry the thing through. Floyd, hearing this news, was hopeful but dubious. He was "sick of the humbuggery" between the Clark sons and the Feils. As far as "the boys" were concerned, he felt there was as much "skin-flint as genius" – maybe a "damned sight more." "It would be a pity," he said, "if love of the ultimate cent would shade that of despatch in a work of this kind." Again, on hearing from the Clarks, he flared, "By Jove, all these fellows seem to have hungered for the whole hog. The devotees of the sublime science are full of 'cents a piece' human nature."[4]

Fraser then called on Edward C. Pickering, the new director of the Harvard Observatory, and hurried out to Lowell to see James B. Francis, chief engineer for the Proprietors of Locks and Canals on the Merrimack River at Lowell, who had visited Mount Hamilton in 1883.[5] Francis was the chief trustee for the Boyden estate. Boyden had died on October 17, 1879, leaving more than $200,000 in trust "to aid in the establishment of an astronomical observatory on a mountain peak at such an elevation as to be free, so far as practicable, from the impediments to accurate observations which occur in the observatories now existing, owing to atmospheric influences."[6] Fraser hoped to persuade Francis to put the telescope on Mount Hamilton. Floyd agreed. But nothing then came of the proposal, and some years later the money went to Harvard.

In the heart of New York City, at the corner of Second Avenue and East Eleventh Street, Fraser went to see Lewis M. Rutherfurd, a wealthy and socially prominent amateur astronomer. A pioneer in astronomical spectroscopy, Rutherfurd had made the first refractor ever designed especially for photography, an $11\frac{1}{4}$-inch instrument, completed in 1864. By this time Floyd and the Lick trustees, as a result of the advances made in astronomical photography by Rutherfurd and by Paul and Prosper Henry in Paris, realized the great advantages that might be gained by the addition to the Lick telescope of the photographic corrector proposed by Holden. If possible, they decided to obtain such a corrector.

After this visit Fraser went on to Princeton. There he found the dome "simply awful," but he was impressed by Charles Young, the director, "a bright young man – the best in Obsy line by far I have ever met. He is a way up in spectroscope work – the best in the U.S." [7] Fraser spent the day with Young and felt he learned a lot. He even talked with "the Professor" about his own ingenious scheme for a $\frac{7}{8}$th-sphere dome for the 36-inch refractor. Young listened with interest and said he was "much in favor of it." But Simon Newcomb, to whom Floyd had written about Fraser's dome, had been afraid "to try so bold an experiment." So the scheme was dropped.

At West Point Fraser called the conditions "terable."[8] He reported to Floyd on the paper dome covering the 12-inch Clark objective. In some places the paper was cracked, and in others it had peeled. Fraser considered paper covering for domes a failure. The shutter, hung inside the dome on a center pivot at the top, swinging around on a track inside the dome, worked badly and leaked. The live ring was all wrong. A great big chain that "cost heaps of money" made the dome revolve around the pier on rails. The 12-inch mounting was rusting. Fraser was disgusted: "I should think Money had been badly wasted here," just as at all the other observatories he had seen. He had expected better of the West Point boys. He felt they spent too much time keeping their brass buttons bright to spare time to keep their instruments clean. He wondered if Holden would have his observatory in good order. If not, he concluded, "we better never turn our observatory over to an astronomer from what I have seen of them."

Thomas E. Fraser's $\frac{7}{8}$th-sphere dome, proposed to the Lick Trust. (Photograph by Loryea and Macauley, San Jose. Reproduced by kind permission of the Mary Lea Shane Archives of the Lick Observatory.)

After observing with Asaph Hall at the Naval Observatory, Fraser had written to Floyd, "He is by long odds the greatest observer in this country. He is a big man in body and mind and you would like him much."[9] By the time he arrived back in California, Fraser felt he had learned enough about observatories to pay for his trip.

While he was away, Floyd, supervising the work on the mountain, was busy "with a thousand things." But he was deeply worried as months passed without news of progress in Paris. He wrote in anguish, "We have suffered a world of anxiety and patience in this matter."[10] When Holden had urged a start on the dome and mounting, Floyd retorted, "The question whether we ought to go on with the construction of the mounting, the turret and the dome for the 36″ before we know that a 36″ is practicable, and before we know its focal length, admitting it to be practicable – is one that has given me great concern and one which I have carefully considered ever since we concluded a contract with the Clarks." [11] He exploded, "I don't propose to turn over to the University of California and science, the big Lick telescope equipped with a 'halo that don't fit.'"

At this time, Floyd had exclaimed also, "When I know the focal length *bang exactly,* I will determine upon the size of the Dome, walls and foundations, the nature of the tower, and mounting, and not until then will I begin construction or to negotiate contracts. I will not make sail to navigate these reefs until I get my bearings." [12]

Nevertheless, Holden continued to bombard him with letters, urging the buying of books, small instruments, and other equipment. In answer to one such onslaught, Floyd wrote from Kono Tayee, where he was laid up with a badly bruised leg caused by a fall from a buggy on the mountain, "I wish I could just have one night's talk with you instead of pushing this infernal pen which I most heartily despise." [13]

By the summer of 1885 a large part of the observatory was finished and in "tip top man-of-war order. Decks swept up and flemished down." Yet Floyd continued to wait for the crown glass. "I shan't do another blessed thing until I know something definite about that matter. 700,000 $ won't last always. We must not 'restez tranquille' about that glass forever." [14] He cabled to Feil, who promised the disks would be ready soon.

Finally, on November 4, 1885, Newcomb cabled the Clarks from Paris the joyful news everyone had been craving for five years. After eighteen attempts Feil at long last had succeeded on the nineteenth: The crown glass was on its way.[15] Floyd, feeling an oppressive weight lifted, prayed that nothing further would happen before it reached the Pacific shore.

The 36-inch disks arrived in Cambridgeport from France, partially ground and polished. But at least a hundred pounds of glass still had to be removed. First a rough grinding was done by machinery. After that, day after day, month after month, the Clarks continued the grinding and polishing, stopping only for frequent testing of the curvature.

On April 14, 1886, Floyd read with interest an article in the *San Francisco Chronicle* reprinted from the *Boston Transcript.* The writer described the crown and flint glasses in the basement of Clark's shop on Henry Street. He called the tables the most costly in the country. If by accident the glasses should be destroyed, he claimed, $25 million could not duplicate them within six months. Although there was always the chance of a mishap, he found everyone optimistic, eager for the day when the lenses, mounted in the "lofty dome" on Mount Hamilton, would become a mecca for "devout" astronomers of all countries. Through this "mighty annihilator of space," he said, the moon would be brought within a hundred miles of the eye of the beholder.

By October the lenses were ready. Floyd asked Newcomb and John Brashear, the optical expert, to examine them. Brashear had observed the great nebula in Vulpecula in telescopes up to 13 inches. He reported now that, if he swept over this nebula with the 36-inch, not knowing *where* he was sweeping, he would never have recognized it. "No Dumb Bell nebula now," he said, calling it a sight of rare beauty. In the pure Mount Hamilton air he foresaw "a grand future" in nebular research with the mighty 36-inch.[16]

When Newcomb went to Cambridgeport to test the 36-inch lens, Asaph Hall accompanied him. They spent a week in Cambridgeport testing the lens. But the tests were hard to make with the temporary telescope mounting provided by the Clarks. It had no clock drive or accurate setting circles. The weather was poor. Nonetheless, on the one night they had fairly good "seeing," they concluded that the lens was acceptable, even though, with the unstable mounting, it was impossible to measure the focal length of the lens exactly.[17] A year later this problem would return to haunt them.

Years before, in 1873, Newcomb had gone to Clark's shop in

Cambridgeport to test the 26-inch glasses for the Naval Observatory. He wrote then of the intense interest he felt in these tests. "The astronomer, in pursuing his work, is not often filled with those emotions which the layman feels when he hears of the wonderful power of the telescope. . . . I was filled with the consciousness that I was looking at the stars through the most powerful telescope that had ever been pointed at the heavens, and wondered what mysteries might be unfolded. The night was of the finest, and I remember sweeping at random, I ran upon what seemed to be a little cluster of stars, so small and faint that it could scarcely be seen in the smaller instruments, yet so distant that the individual stars eluded even the power of this instrument." It was impossible to determine the position of this cluster, because he had no clock and no circles. Newcomb concluded, "I could not help the vain longing which one must sometimes feel under such circumstances, to know what beings might live on planets belonging to what, from an earthly point of view, seemed to be a little colony on the border of creation itself." [18] Now, years later, testing the far greater 36-inch lenses, Newcomb must have wondered again what mysteries would be unfolded by this, the greatest telescope in the world.

Soon after this Fraser headed east to bring the objective to California. In Cambridgeport, gazing on the precious lenses set in their cells, he exclaimed, "Their impression on me at first sight was grand and everlasting. I beheld for the first time one of the greatest things in science. I also saw before me the master stroke of genius that had made for the Lick Observatory what the donor had commanded, the largest objective in the world. I knew now that the Lick Observatory would be an assured fact. All these thoughts rushed through my mind and I shook hands with the old gent as he came in. I said, 'You have done well and have given us for Cala something that cannot be bought with money.'"[19] Shortly afterward the news came from Mantois in Paris that Charles Feil was dead – a victim, perhaps, of the long agony over the crown glass.[20]

In Cambridgeport Fraser began packing the glasses for shipping, treating each component with great care, wrapping it in layers of flannel, then placing it in an inner box with curled hair, and finally in an outer box with coiled springs. On December 18 the boxes were loaded on a sleigh and taken to the private railway car *David Garrick.* There they were buried in mattresses. Fraser joined them on their long journey – to San Jose, by way of New Orleans and El Paso.

Waiting in California, Floyd confided his concern to Burnham: "I most long for the relief from anxiety which its safe arrival on the top of this solid rock will give us all." [21]

Arriving home the day after Christmas, Fraser wrote exuberantly, "We are in California all right, all things safe." [22] At the San Jose depot Fraser was met by Floyd and Mastick. The Captain stayed in the car all night. Before daylight the following morning, they loaded the boxes on an especially arranged spring wagon for the journey to their "future field of Research." It was a beautiful day as they trotted up "the smooth and splendid road" to the top. They arrived there soon after noon. Mathews was waiting to take a photograph of the arrival of the 36-inch objective. Floyd and Fraser unpacked the lenses, and with intense relief they saw that all was well. For safekeeping they locked the glasses in the vault of the North Dome, to wait there until the mounting and dome were ready.

In the summer of 1885 a California senator and university regent, Judge John Hager, had proposed that Edward Holden be made president of the University of California and then director of Lick Observatory when it was finished.[23] Floyd, who considered Holden the best choice for director, was enthusiastic. He felt it would be good to have a president of the university with a brain untrammeled by prejudice. "That is not to be found in the gray-headed old groves of Grekko-theological Latin fossils." He felt that the university was in the doldrums. "Prayers and dead language chatter wont move it ahead." He thought it needed a shove from somebody who was not burned out. He was sure Holden could leave it in "smooth water and fair breezes." [24] At this point he looked forward to Holden's coming. Yet, when Holden suggested taking on both jobs simultaneously, Floyd, after meeting

Arrival of the 36-inch objective on Mount Hamilton, with R. S. Floyd *(left)* and T. E. Fraser *(right)*. (Reproduced by kind permission of the Mary Lea Shane Archives of the Lick Observatory.)

with J. H. C. Bonté, secretary of the regents, told the astronomer that would be impossible. "The Regents wont pay a cent for an astronomer until the Observatory is turned over to them – and some of them think its going to be an elephant."[25] One of the regents was George Davidson.

After hearing of the recommendation, Holden wrote to Hager to press his claims.[26] "You probably know that the plans from which it [the observatory] is built are mainly by myself, that I have ordered and mounted all its instruments and that I am familiar with every detail of its construction."

Holden, who had been craving the post for nearly ten years, had told Floyd, "I certainly should like to direct your observatory, but I don't belong to a hungry pack; and I respect my

Richard S. Floyd with the 36-inch lens on Mount Hamilton, 1881. (Reproduced by kind permission of the Mary Lea Shane Archives of the Lick Observatory.)

friends and most of all, you, my dear Captain who have been more than kind throughout."[27]

From Kono Tayee Floyd wrote warmly, "I wish you were here now to fish with me – splendid way to talk free from all interruption except the pleasant one of a bite."[28] Cora had just returned from hunting, bringing back a springtail and a "very fat" canvasback. What a banquet they could have with such fare and some splendid old Mouton-Rothschild, Château Margaux, or Château Lafitte, vintage 1869, that he had bought in France and shipped around the Horn!

Floyd offered Holden perfect freedom with a passkey, a room, and desk of his own. "You can dine out and sleep out without notice. . . . You may be sure that I will with the most watchful care see that not one of the nonsenses or the conventionalities of society shall embarrass you in the lots of flying about that you will perhaps have to do in the process of getting settled in your new duties."[29] He concluded in nautical lingo, "And now my dear Boy I hope that I am congratulating you on the sheer pole of the rigging that leads aloft to fame–and that I may have the pleasure of sending you many a hearty one up the ratlines, as you skim over the futtock shrouds, in the topmost cross trees–and finally when you can wave your hat from the lightening rod above the main truck." He added, "I have a million things to talk to you about," and signed, "With love, your friend, R. S. Floyd."

When Newcomb heard of Holden's appointment to the presidency, he was more cautious. At the end of February 1885, he had confided to Floyd his choice as Lick director and had written, "If I were to fill it tomorrow, I should say Todd without asking any further questions. This remark is of course private and personal." Now he advised Holden to be "as wise as a serpent though it will hardly do to be quite as harmless as a dove." He had one wish, he added, "respecting your policy to wit: that you will keep out of all fights, except those you are sure of winning." [30] Over the years he had come to know Holden better and had apparently become somewhat disillusioned.

When Holden left the Washburn Observatory for California, his wife, Mary Chauvenet, moved back to St. Louis to live with her mother. Their son Ned was in boarding school but would spend time with his father on Mount Hamilton. In those days such a marital separation must have been frowned on by the faculty of which Holden was about to become president.

In January 1886 Holden arrived to take over the presidency. He lived first at the Union League Club, then moved to 908 Pine Street in San Francisco. Floyd welcomed him there and invited him to see *Othello.* "I want to enjoy the play with you," he said. And again, inviting him to take potluck, he asked the astronomer to hear a performance by Alessandro Salvini for which he had just procured tickets. "Don't wear full dress. Nobody will." [31]

In June Holden was formally inaugurated president. As time passed Floyd had become increasingly worried about the future of the observatory itself. He feared that, however successful the construction might be, it might fail for want of an assured and good salary for the director. This salary had to come from the balance left after the observatory was finished. With the great

Edward Singleton Holden, first director of the Lick Observatory. From a photograph taken in 1886. (Reproduced by kind permission of the Mary Lea Shane Archives of the Lick Observatory.)

instrument a success, a first-class director could get anything within reason from the legislature, not for his salary, but for needed repairs, reasonable improvements, extra instruments for special phenomena, extra lamps, telegraph keys, and staples. But he had to be in a position to put his requests independently and in the most advantageous shape. "To a hungry half shell vegetarian-looking representative of a sublime science who walks to Sacramento and BEGS for things – and begs first withall for his living, nothing will be given."

8

Engineering feat on Mount Hamilton

In 1884 the meridian circle was ordered from Repsold in Hanover, Germany. For it, Alvan Clark made three 6-inch objective lenses. One lens was the permanent objective of the instrument, one was a permanent lens for the collimation mechanism, and the third was a dual-purpose lens for occasional recollimation and for use as the objective of a portable refracting telescope.[1] Of these the last was the most important in the history of the 36-inch telescope and also in that of Warner and Swasey. With it the partners got a chance to demonstrate their engineering and instrumental skills and to prove to the Lick trustees their ability to handle the huge mounting for the 36-inch. They felt, perhaps mistakenly, that such a telescope on a small scale could be replicated on the much larger 36-inch scale.

Warner and Swasey claimed that in the 6-inch telescope they dispensed with the ropes and cords leading from the mounting head to the eye-end controls usually used to move and clamp the instrument. Instead they used only two handles at the eye end, one for clamping and slow motion in declination and the other for clamping and slow motion in right ascension. In place of ropes they used parallel rods running along the tube. "The rods acted through gearing to move the telescope in slow motion and to tighten and loosen the clamps. Each rod was made in two concentric parts – a hollow tube and an inner solid rod." To illustrate their ideas they sent along a sketch.[2]

It is always hard to establish priority in scientific invention.

Henry Draper 12-inch telescope on Mount Hamilton. (Reproduced by kind permission of the Mary Lea Shane Archives of the Lick Observatory.)

Grubb claimed he had been the first to use eye-end controls in the Vienna telescope, whereas Fauth, working with George Saegmüller in Washington, had pioneered in the field on a smaller scale. Warner had spent hours at the Philadelphia Exposition studying Fauth's methods, and he had seen Grubb at work in Dublin. If, however, he and Swasey were not the first to build a mounting with eye-end controls, they certainly perfected the method. The observer using periscopes could now read the graduated circles either from the eye end or from a control station on the pier of the mounting. Warner and Swasey had also improved on Alvan Clark's driving clock, using better counterweights. In all their work they showed their ability to produce the highly precise instruments needed to make the most delicate, intricate astronomical measurements. In the 9½-inch refractor for the Hartford Public High School, in Connecticut, as they told Floyd, they had their first chance to build into a telescope all their basic

Sir Howard Grubb with 6½-inch portable equatorial, 1877. "All clamps, slow motions and Reading of R.A. and Decl. available from eye end. Howard Grubb, Dublin, 1877." (Courtesy David DeVorkin, National Air and Space Museum.)

design improvements in equatorial mountings and to show that they were talented artists as well as skillful engineers.[3]

At this point Holden wrote to Floyd asking if he could negotiate the contract for the 6-inch refractor at the Lick Observatory, incorporating his design ideas.[4] He specified that the mounting had to be steady and adjustable with a range of latitude from fifty to ten degrees. He agreed overoptimistically with Warner and Swasey that there was nothing more or different to be done with the 36-inch than with a 6- or 12-inch. Burnham concurred. "There is certainly no instrument where extreme simplicity in mechanical matters is more necessary than in an equatorial."[5]

That spring, after inspecting the Leander McCormick Observatory in Virginia, Fraser went with Holden to see the new 6-inch mounting in Cleveland, where Warner and Swasey had moved their shops late in 1881. Holden wrote enthusiastically, "If the final telescope were to end up as pretty as the blueprint, you will have the handsomest mounting in the world."[6] He was particularly pleased with the slow motion and clamping controls that were similar to the Hartford telescope. Fraser was also enthusiastic. "I have got Warner and Swasey down to a line as to how they will proceed to Draw Plans and what they will doe. . . . They are very smart and Swasey a sharper and smart as any man that walks."[7] He had a hard time holding his own against the young engineer. On the other hand, he found Warner "not so smart."

So it was that Warner and Swasey, dreaming of a magnificent future, began to plan for the great telescope even before the 6-inch was finished. While arranging to ship the small mounting, they said they intended to submit designs and bid on the 36-inch. For that mounting Floyd could judge their design concepts, because they had incorporated the principal design elements of the 36-inch in the smaller prototype. They wrote confidently, "We have the motions worked out, in much better style than we expected would be possible, in fact, we have greatly improved our former methods of solving the most difficult problems."[8] They repeated that they were sure they could do on a large scale what they had done so well on a small scale. It would only be a question of engineering skill instead of just instrument making.

Three years earlier, in an extravagant letter to Newcomb, Floyd had suggested that the mounting could be built in six months, not three years, if adequate time had been spent on the design. "Given enough money (say $20,000,000), and I would like to undertake to do the mounting in Frisco within six months for an Equatorial from 100 feet to 100 yards long. . . . When we let the great shops which build steamships, locomotives and mining machinery know that optical instruments have gotten up to their scale you will see that all the old fogy, stay in the groove, put-you-off, talk from the point fellows–like the Grubbs, Repsolds, Clarks etc. will have to redesign the heavy machinery. We are just now arriving where astronomical instruments of great size are getting sufficiently in demand to require foundries and great machine shops for their heavy work instead of the simple facilities required to produce the mountings of opera glasses and microscopes."[9]

Receiving these and other letters in similar vein, the conservative Newcomb, put off by Floyd's flamboyant and prophetic statements, answered cautiously. He too had often thought that, with the help of a first-class designer of big machinery, he could make as good plans as anyone. But he said he hardly dared take on the responsibility. To Holden he confessed he found Floyd's letters disquieting, "a mixture of badinage in what he says which tends to conceal his real intentions or serious views, if he has any yet matured."[10]

In an article for *Harper's Magazine,* Newcomb elaborated on the problems of building a mounting that, in size and weight, was a very heavy piece of machinery.[11] As Floyd proposed, it could be built in a shop devoted to the construction of engines of the largest size. Yet although it would be strong, the masses of metal forming its axes and supports had to be moved by a mechanism as delicate as that of a watch. These and other instrumental problems required the combined skill of the astronomer, the astronomical mechanic, and the engineer.

So the time came to open the bidding. The Lick trustees sent out requests for bids to five firms–two in the United States, one each in Ireland, France, and Germany.[12] While Warner and Swasey felt confident they would be awarded the contract, Howard Grubb in Dublin was still hoping for the chance to build the equatorial and wanted to be sure his estimates would be given equal weight with those of other firms. The same, he hoped, would be true for the dome. He pointed out that the 45-foot dome for the Vienna equatorial was a "perfect success."[13] If given the chance, he could send out such a dome by the following summer.

At a meeting of the Royal Astronomical Society in London, Grubb demonstrated his model of the 36-inch mounting designed in the competition for the Lick trustees. He said, "The idea I have had all through is to bring under the direct control of the observer all the required motions of the instrument and of the dome, so as to give him as little physical exertion as possible."[14] He showed how he had arranged four small water engines, one to give motion in right ascension, another in declination, a third to move the dome, and the fourth to bring the observer into proper position for observing. He planned to control these engines with an electrical apparatus the observer could carry around with him. In his model he provided various keys; one would turn the dome in one direction, the other would reverse it; another would light the observatory, and one of the most important keys would make the whole floor move up or down to put the observer in a good position for observing.

The president of the society, impressed by Grubb's model, envied the luxuries to be provided to future astronomers. "It seems as if the life of an astronomer, with simply a key in his hand to touch in order to make these movements, would be something that one could dream of but could scarcely hope to realize." Another member noted facetiously, "We may suppose that the door moves with the floor, otherwise a gentleman might go in after dinner, and the floor might be down 30 or 40 feet, and then the observation would be spoiled."[15] Still another mourned, "All I can say is one must wish to go over life again to have such an instrument."[16]

As for the Lick contract, Grubb pointed out that three months earlier he had been told that the estimates for the equatorial had to be in San Francisco by May 5, with the dome complete and in working order by October. To accomplish this, he said he would have to turn months into years. On July 8, 1886, Mathews, in a letter that amazed Grubb, explained that the Lick trustees, "with a compelled special consideration of the element of time, which circumstances now make one of vital interest to the work," had let the contract for the mounting to Warner and Swasey for $43,000 to be delivered complete at the observatory on Mount Hamilton. Although it was the highest bid, they considered it the best. They felt, too, that such a huge mounting could be built in the United States with the least amount of "humbug." The contract for the steel dome had gone to the Union Iron Works of San Francisco for $56,850 "under the same circumstances." Mathews thanked Grubb for "the prompt and courteous manner" in which he had responded and assured him of their appreciation of the very great disadvantage "to which you have been put by the remote situation of your works from California and also the very short time for consideration that circumstances permitted them to offer you." [17]

In awarding the contracts Newcomb's original scheme for bidding had again been adopted.[18] As a result Grubb received a paltry $600, of which $400 was for the use of his plans and specifications for the mounting and $200 for his plans and specifications for the dome "according to the terms of the letters to you from the President of the Trustees dated respectively Feby 2/86 and Mch 1/86." [19] Grubb was naturally profoundly distressed by such a scant return on all his years of work.

Meanwhile, in Cleveland, Warner and Swasey were going ahead with the pier design. It was cast in sections so that it could be shipped by rail across the country, then hauled up the tortuous Mount Hamilton road. Floyd had suggested the places where the breaks in the pier should be made.

For a time work had to be delayed because the drawings of the elevating floor, originally designed by Grubb but modified by the Union Iron Works, had not been completed. Without these drawings, Warner and Swasey could not design the spiral staircase along the south face of the pier, as they had to know the exact dimensions of the central hole in the movable floor. Nor could they produce a final design for the pier and mounting head where certain controls were to be located. Four months passed before these problems were resolved.

When the drawings finally came, Floyd found that the shape of the entire pier had been altered so that it would better fit the hole. Originally the pier had a gradual, graceful curve. Now it was nearly rectangular at the top. Floyd, with his keen artistic sense, was disappointed. He urged the partners to keep part of the original curve near the base where the floor, at its lowest position, could not interfere.

"For more practical reasons" Floyd proposed other changes. One of these was for hand holds in the pier, to which Warner and Swasey objected. Floyd was incredulous. The question of looks, he noted, "is of little moment considering that we will possibly have to cover the entire pier with a wooden jacket to avoid the effect of changing temperature." Instead he offered to fill the oval openings with "handsome wood or plate glass." [20]

Floyd found too that Warner and Swasey were having trouble with the casting of the iron column sections at a local foundry. The large base of the pier caused trouble. More serious trouble came with the huge polar axis that had to carry the brunt of the thrust of the great instrument. They found high-quality carbon steel hard to get and had to cast the polar axis three times before achieving success. They had hoped to ship the pier and mounting in June 1887, but this proved impossible.

Meanwhile they were working on the mechanical components of the equatorial head. According to Edward Pershey, "The gearing and power drive mechanisms of their new design were the first fundamental alterations in telescope control since the early part of the 19th century. It was their machine design creativity based on the construction of production machine tools, that allowed them to be innovative as instrument makers." [21] He emphasizes that, in order to use the solid rods and gear trains, instead of dangling ropes, they had to produce gearing and bearings of exceptional quality. Only in this way could a single observer at the eye end of the telescope produce movement by simply turning a reasonably sized knob and wheel. Only then could the observer focus on a star or other heavenly body with the requisite accuracy.

As the work went on, the strain was felt in Cleveland. In August 1887 Swasey, suffering from exhaustion, was forced to take time off. The load then fell on Warner, who admitted that Swasey had been responsible for most of the best ideas on the huge mounting.[22]

In late August Burnham arrived in Cleveland to inspect the progress. He was impressed by the drawings – more than 180 in number – and reported to Holden, "It struck me that he [Warner] has worked out . . . matters of detail which have come up in the course of the work, with great ingenuity."[23]

The drawings included those for the telescope tube, which raised other problems. It would be rolled from Pittsburgh plate steel, but even the best steel flexed as the tube rotated along the two axes of the mounting. The tube had to be rigid enough to maintain optical alignment in every possible position, and thus a simple cylinder would be impossible. Therefore, even before the partners were awarded the contract, Holden had proposed a narrower, more tapering tube, reducing it from 48 to 42 inches. In the end Warner and Swasey, after making further calculations on the flexure, returned to the original 48-inch diameter at the objective end, with 36 inches at the eye end. To check their calculations tests were made on the deflection of the tube. The tube passed the test "with tolerances expected for the optical system."[24]

Floyd subsequently came up with an ingenious idea. He suggested that "in order to keep the flexure a known constant, the maximum weight of all possible pieces of equipment be loaded onto the tube from the beginning." As different attachments were added, the equivalent amount was removed from the dead weights. In this way the flexure was kept as a constant and eliminated the need of sliding weights and readjustment of the balance of the instrument each time a different attachment was being used. "This was an innovation in astronomical instrumentation."[25]

Before the final contract was awarded, Warner and Swasey had submitted three different tube designs to Holden, who had passed them on to Floyd, asking him to make the decision because he "could not care less."[26]

Holden's comment and his professed lack of interest in the telescope tube may have had its source in an altercation with

Floyd sometime before. Holden had evidently given Warner and Swasey the idea that he was in complete charge of the mounting, including the awarding of the contract. He reaffirmed this idea when, on his own, he arranged a contract with Warner. Floyd learned of the agreement when Warner arrived in San Francisco. He was not only shocked but furious. He had never given Holden authority to act in this way for the Lick trustees, and he was quick to let Warner know his feelings.[27] Warner had gone to Salt Lake City, thinking everything was settled. He was forced to return to San Francisco. When Floyd then went over in detail the drawings submitted by Warner and Swasey, he was able to reduce the bid for the mounting by $8,000. Later that same month the new contract for $45,000 was signed by the Lick trustees for the complete mounting to be delivered at the top of Mount Hamilton.

To Fraser, Holden's actions at this time confirmed a gradually growing sense of the astronomer's duplicity. Looking back in 1888 he wrote to the Captain, "I told you of him a long time ago when he wrote that letter before he came out, as to who should superintend the Dome, and then on the bad side of it how was it that Warner should go to Salt Lake with $8000 more for his contract than what he got for it afterwards from you." "Where," Fraser asked, "was the money going?" He added, "You remember when I told you about the copper mine proposition and you got so angery at me for hinting such a thing. These kind of things looked strange to me years ago, he has made his point he is established for a while but let him look out the men that has the name of the work done at the Lick Observatory cannot be hurt by him. It will onely be for a while if he keeps on with such statements as he is now making he will have to step down and out."[28]

Floyd's doubts about Holden were also growing. When trustee Plum returned to California from the East, Floyd confided his own uneasiness over Holden's relationship with Warner and Swasey. "They have been very derelict in keeping me advised, and I am inclined to think they must be under the impression that they are making the mounting to please one astronomer only and that the Lick Trustees are but figure heads in this matter. I should be very sorry if we should be compelled to disabuse their minds at the expense of delay. I hope I am wrong."[29]

Soon after Holden arrived in California, Floyd had written to him about the dome and his impatience to start work: "We've now lost a bang month since you came I have a good notion to order the Dome immediately and be done with it. I can start the whole thing full blast in a week and finish it in 5 months. Chaw over this until I see you – may be tomorrow – Saturday I can't – Sunday if possible – certainly Monday. Don't forget you are to take pot luck with us Monday at 6."[30]

In 1883 Floyd had begged Newcomb, "Tell me all you can about Domes! . . . Don't forget to write me everything you have observed about Domes – that's the subject I am at present crazy about. How much do they weigh? What kind of rolling gear? What kind of turning gear? How many pounds to start them? How many to keep them going?"[31]

At the University of Wisconsin, Holden had hired a young engineer, Storm Bull, to make the computations for the strength of the arches and walls of the dome. Fraser and Floyd had studied Bull's strain sheets, and Fraser concluded that although Bull might know what was required in the East, he had not the slightest idea what was needed on Mount Hamilton. He remonstrated with Floyd on allowing Bull to go ahead with work that was of no earthly use to them. Later he claimed money had been thrown away, as none of Bull's work was ever helpful in the construction of buildings and dome. Bull had wanted to add buttresses to the circular brick wall on which the dome would rest. Floyd argued that these would only increase the expense and cause inconvenience, blocking the scant roadway around the tower. They would "catch more wind, shy dust into the windows – increase vibrations and make the whole thing look like a synagogue."[32]

Comparing Storm Bull with Warner and Swasey, Fraser counseled Floyd, "The fine theory of a well Educated Man in the

Science of Mechanics without actual Experience in construction interfere with the mechanical ingenuity of such men as Warner and Swasey."[33] He approved the young engineers as "practical men."

In answer to a flood of letters on Bull's computations of strength, strains, and pressures for the brick tower to carry the large dome, Floyd pointed out that "such an important part of the data in Mr. Bull's calculations is assumed – and assumed from mere guess works – that the conclusion seems to me only a kind of random shot and nothing assured save that it will hit some where at a large and definite distance on the side of safety."[34] Fraser, calling the proposal dangerous, was even more vehement. "A man that can consider a force of 50 lbs. of wind pressure to the inch must be Storm by nature as well as by name." He thought buttresses on a dome as absurd as making the smokestack of a steamship square. The dome was a buttress in itself. The gusts of wind would be better resisted in a round smooth dome than one with projections and twenty feet between supports. He called the buttresses a humbug and concluded, "They are not required here."[35]

It was hard to solve these problems through correspondence. Before Holden arrived in California, Floyd had hoped to go East to discuss them with him in Wisconsin and Newcomb in Washington. But Cora would have none of it. He had written to Holden, "I am in the glooms about my proposed trip to see you. . . . I broached the subject yesterday and developed a regular Wisconsin tornado. I buttressed myself up bravely, upheld my proposal with all the components of Mr. Bull's report, strained around about untapered girders at Cleveland, and figured out an absolute necessity. But she couldn't see why!"[36]

Floyd hated to give up the trip. But in the end, he and Fraser had their way, demolishing not only the buttresses proposed by Bull but also his dome design that Holden had advocated. When Holden urged that the dome be made even larger than Floyd proposed, in order to accommodate the eye end of the spectroscope, Floyd countered, "If you want to look at the Zenith you can have a hatch in the deck at foot of pier and look aloft from cellar."[37] Anything, he felt, would be better than increasing the dome size at enormous extra cost.

From the beginning the Lick trustees had wanted the work on the Lick dome done in the United States – if possible, in California. They knew that the Captain shared their view and "was all for California," provided it was possible to get as good a design there as in other places.[38] They believed too that this would have been the founder's wish. Floyd echoed their feeling when he wrote of the work of the European firm of Repsold, after seeing what the Clarks, Warner and Swasey, and the Union Iron Works could do, "We can certainly get as fine work done in this country as the Repsolds can do with great chances of better success. There are oceans of the finest mechanical skill developed in this country in many directions which if turned on a fine meridian circle would I really believe produce the best one ever made. Any improvement or new idea that it would take the Repsolds a year to have dawn about them would be in full daylight over here in a minute. They rest upon the reputation of the best thing for some years and take improvement like medicine.

"Of course," he went on, "I don't believe that national prejudice should influence such things but I don't think we should go abroad when we can do as well, with strong chances of better at home and especially in the case of our National Government, which can well afford experiments of this kind to induce the over abundant and ambitious skill of its citizens towards furnishing its scientists with instruments."[39]

For these and other reasons, including the pressure of time, Floyd and Fraser had decided that it would be ridiculous to go East for the dome if Irving M. Scott at the Union Iron Works in San Francisco would give them a fair price. Scott, an engineer, had first designed machinery for the Comstock mines in Nevada. In 1883, under Scott, the Iron Works had shifted from mining equipment to ship building in the Potrero district of San Francisco near Hunter's Point. Soon Fraser was meeting with Scott and George and James Dickie who, before joining the firm in

1883, had built wooden ships in Scotland, then had turned to the building of iron and steel ships for commercial and naval use. The Union Iron Works, builders of mining machinery since the 1849 gold rush, in 1876 had built for Pacific Mail the *City of Peking,* an iron screw steamer to sail between San Francisco and Hong Kong. In 1885 the *Arago,* an 800-ton steam collier, the first steel steamship on the Pacific Coast, was launched, and Floyd was there to celebrate. Afterward he bragged patriotically about this ship, "built complete at our new Iron and Steel Works which in its convenient arrangements, splendid machine shops, modern tools etc. etc." put anything in the East in the shade. "We now build the largest steam engines in the United States. We are ready to help build our new Navy and are ready to take a contract for a 15,000-ton turreted iron clad immediately. Domes, tubes, polar axes etc. are mere bagatelles."[40] In January Floyd joined Scott at the Union Iron Works to discuss estimates on a steel dome 75 feet 4 inches in diameter on the outside, 71 feet inside. The dome was based on Fraser's composite design, its parts selected from the diverse domes he had examined at the Washington, Dearborn, Washburn, Annapolis, and Harvard observatories, as well as information gained by Newcomb and Floyd in their European travels.

As they went ahead with the final dome design, Floyd had asked Fraser to prepare plans and specifications that would not infringe on any patents covered by Warner and Swasey. Fraser sent to Washington for all patents on domes, and went to work to draw the great dome on paper. He worked for three months on the design. If visitors distracted him during the day, he worked at night, perfecting the design, then going over it with Floyd. They were then ready to go to George Dickie and the Union Iron Works.[41] There a mechanical draftsman would work out the final set of drawings and specifications "that beat anything in that line ever done." This took over two months. Changes were then made until everyone was satisfied that "there would be no plans received by the Lick Trustees to compare with them." For legal reasons they were also approved by James T. Boyd, lawyer for the Lick Trust. This was done despite the fact that no one at the Union Iron Works had ever been in a large observatory. Floyd pointed out afterward that one of the most important things in the dome design was to allow for the expansion and contraction of the iron plate on top of the brick wall. The outside circumference of the iron plate was about 235 feet, and differences in temperature could make this ring shorten and lengthen about $1\frac{1}{2}$ inches.

On May 13, 1886, Floyd signed a contract with the Union Iron Works for the giant steel dome.[42] The weight of the dome's moving part was estimated at 142,535 pounds, the live ring at 26,052, and the lower rail and base plate at 75,250 pounds. As in the 12-inch dome, an endless steel wire rope would be used to revolve the dome. It was agreed that the workmen who would mount the dome on the "hill" would be housed there and that the contractor would provide food and cooking for them. Time, Floyd emphasized, was of the essence.

Time was indeed of the essence, and this again weighed heavily against Howard Grubb in Dublin, as it had in the case of the 36-inch mounting and the 12-inch dome. Ever since they met in Dublin, Floyd had valued Grubb's suggestions for the design of the telescope and dome, and especially for the elevating floor. Therefore he now felt guilty about awarding the contract to another firm. He had asked Holden, "Don't you think we ought to ask Grubb to propose for our *Dome* at least? He has certainly had more experience about large Domes than any other manufacturer."[43] He added, "The rolling gear and every thing nearly that's good about what Warner and Swasey have done in this direction is borrowed directly from Grubb. There is very little if any thing original in the mechanics of Mr. Grubb's Domes and nothing whatever in Warner and Swasey's, but Grubb was first to apply many things already known with advantage to Astronomical Domes – and as he has the most experience with big moving Domes of anybody – except now Struve and the Nice people. Warner and Swasey made Grubb's rolling gear for our 12-inch Dome at our dictation and did a very fair job – nothing extra.

Sir Howard Grubb. (Reproduced by kind permission of the Mary Lea Shane Archives of the Lick Observatory.)

They made the same thing without improvement except perhaps in workmanship for the Virginia Dome, and had to resort to the most old fogy and stupidest methods for moving it." "Nor," said Floyd, "is there any original principle in the shutter that W. & S.

claims as theirs and have a patent on. It's a bad thing to seek a patent on this kind of thing."[44]

Mathews, speaking for the Lick trustees, emphasized in a letter to the *Observatory* that the trustees had paid Sir Howard Grubb for his ingenious idea on the lifting floor.[45] The *Observatory* editors agreed that the trustees might be within their legal rights, but they argued that there was another side to the transaction by which Grubb "had been considerably aggrieved." The editors intervened on scientific grounds. "The facts are these," they explained, as they told of the trustees' contractual arrangements on the telescope and dome. "Such a method of dealing with a competitive trial may not be unknown in very sharp business practice; but we say emphatically that it does no honour to men who are entrusted with a large sum of money generously bestowed for the advancement of science. No one will deny the claims of the great instrument makers to rank as men of science; and in dealing such a direct blow to scientific enterprise, the Lick Trustees were acting in direct opposition to the spirit of the Trust they represent."[46]

Despite this disappointment Grubb wrote warmly to Holden, congratulating him on the magnificent fields of work before him and wishing him every success.

In the spring of 1886 Floyd and his family moved to the mountain to live in the three-story red-brick building known as the "astronomer's dwelling." The decision was not easy. Cora understood his reasons, but she hated to be away from Kono Tayee. The months, especially in winter, were long, and life, she felt, would be hard. As it turned out, the winters of 1886 and 1887 proved to be exceptionally cold, and the mountain was often covered with snow. To the amazement of its inhabitants, four inches of snow fell, even in San Francisco, in February 1887. The Floyds moved into the "dwelling" where single astronomers would one day live.

That summer a Mr. Tison, a friend of Floyd's cousin, Marmaduke Hamilton, arrived on Mount Hamilton. Approaching the

Henry E. Mathews in the library on Mount Hamilton, "a handsome room finished with white polished ash book cases and fireplaces for ventilation." (Reproduced by kind permission of the Mary Lea Shane Archives of the Lick Observatory.)

summit he compared it to a Scottish castle. At the door, as Tison jumped from the buggy, Floyd met him with a hearty grasp. Tison wrote afterward, "There he stood, a born soldier and gentleman, a picture of perfect health and contentment. From that moment, he and his household were unceasing in their hospitality and when the time came the next day to leave, there was that which will not be effaced from the bright side of things." [47]

Later, when asked to pay board for his family, Floyd protested to Mathews. He noted that because of his responsibilities and the unexpected delays that dragged into 1888, he was *compelled* to keep his family on the mountain, "much to their inconvenience and discomfort, and contrary to my private arrangements." [48]

His family included Cora Floyd, her niece, Cora Matthews, and thirteen-year-old Harry. It was in a way an odd group, forced by circumstance to spend months in virtual isolation. The Captain was intensely busy with Fraser, rushing to finish work on the observatory. He had less time to worry about the details of daily living than the others. In good weather Cora Floyd went hunting or fishing. When it was bad, she could read or play the piano Dick had brought up the mountain for her. At night there were often lively poker games. For Cora Matthews, a charming young girl accustomed to plantation life at Oakley, this existence must often have seemed strange although she soon discovered a close friend in young James Edward Keeler, who had arrived on Mount Hamilton in 1886. During the day she usually tutored the shy, artistic Harry, who spent much of her time in a world of fantasy.

A lover of romantic novels, especially those of Alexandre Dumas, Harry was never happier than when she was enacting the roles of the characters in her books. At other times Harry entered the real world. As a tomboy, wearing knickers, she had "lots of fun" riding over the mountain on horseback. Like her father and mother, she prized her independence.

Like his daughter, Floyd longed for the day when he could do as he pleased, when he pleased. One day he wrote wearily to Holden, "I wish the Observatory was finished and you had it. I have several things that I wish to devote myself to on my own hook – where I can be entirely independent of the public – and where I can just revel in mistakes if I choose to – as a kind of luxury – and without making the public or anybody unhappy."[49]

But for many months, Floyd was shackled to his job on the mountain. Often, in their isolation, cut off from the world below, the Floyd family lacked vital needs. One day he pleaded with Mathews, "For God's sake send me up the latest Directory – lots of stamps and stamped envelopes. Am completely wrecked and stranded for want of them with crowds of important correspondence. Hurry up! Right off! Express! The last stamp goes on this. Expect to hear from me no more until you save us from ruin with the wherewithal of communication."[50] As the kerosene supply ran low, Floyd cried, "I am afraid we will be clean out and in darkness before I get another waggon up the road."[51]

Often, too, their food supplies ran low and had to be stretched to provide for strangers who, after climbing the mountain, found it too late to return home. Fraser recorded some of these in his "log":

> June 23, 1885 – Stranger Supper Lodging Bed and Dinner $2.00.
>
> July 12, 1885 – Three ladies lodge all night.
>
> July 15, 1885 – To Rich and Party of 5 Supper & Bed. To one bottle of claret furnished by Fraser 50 c.
>
> Aug. 16 – Judge Shaw and a party for Dinner & Breakfast and Rooms & for 3 horses $9.00.
>
> April 11, 1886 – Mr. Start & a Party of Seven all night Caught in a storm and had their own provisions $7.00.[52]

Other records show that campers were fed, "put up in the hay," and given coffee.

On Tuesday, June 15, 1886, Fraser arrived from San Francisco, leaving San Jose at 4:30 A.M. after visiting the Union Iron Works and inspecting the "work now being got out at their great workshops in S.F." "The patterns for the plate and the Castings for the plate to rest on is allout and ready for the plainer. There was some changes in the shutters I brought up the plan of them for Capt. Floyd to pass on and also to consult about changes with Prof Holden and myself I got on the Hill at 10 AM and laid out my work for the Brick Layers they arrived at 4 p m direct from S.F. Mr. Piercie and 5 Brick layers."[53]

This led Fraser to another celebration – the laying of the cornerstone for the great dome on June 16, 1886. The captain, Cora, and Harry were on hand, together with Tom Fraser and his wife, James Keeler, Edward Holden, and Cora Matthews, along with the workmen, including five bricklayers and two hod carriers. There, too, were Sirius and Procyon, the only two Ulner hounds in California, which Holden had just given to Floyd. Harry laid the first brick. Fraser recorded his sentiments in his "log": "As the sun was setting in the western sky we all gathered around to witness the laying of the First Brick in the Radial line of the dome, a ½ Dollar which date is the anniversary of our national existence and the Death of the Benefactor Mr. James Lick who gave this Institution for the Advancement of universal science." He continued in a mystical vein that would have pleased the benefactor: "The hour was late and in accord with the silent stillness of the evening shades of the coming night. For inside these walls in the time to come silent watchers will softely tread and from the setting Sun till the Rising of the same Astronomers and scientists of every scool and Race will here within these walls congregate to learn of all marvelous things of space; according to the wishes of the President of the Trustees there was to be no Flowery speeches so it was imposable for me to get in with a set speech in memory of this great event in the History of Science." Fraser must have been deeply disappointed, although he realized it was hard to do the subject justice "and silence was the best then." Floyd's idea "was that it ill became us in rearing up this monument, the greatest at this day to science, to spread ourselves like politicians in grand oratory on a subject so little known, and it was in profound silence we commenced the erection of those walls that might in the future eclipse all the powers of oratory.[54] A leader in that eclipse would be the twenty-eight-year-old Keeler, who had arrived on Mount Hamilton to help set up various instruments and establish a time service for the West Coast.

Holden, a stickler for accurate time, was enthusiastic about the program. "Every train west of the Rocky Mountains runs more safely for the observations which are nightly made at Mt. Hamilton and every shipmaster who will take the trouble, can make a better landfall at the end of a long voyage because the observatory has fine instruments, good clocks and faithful observers."[55]

At first Fraser, dubious about Keeler, was afraid Floyd had done the wrong thing by allowing an astronomer on the mountain, where he might interfere with the building. But he was willing to wait and see. "It will be all in the man. If he has the right ring, all will be right, but if stubborn, then things will go wrong and he will have to leave. . . . If he turns out all right I will be his friend, but if he dont I have no use for Drones here."[56] He was afraid, too, that Keeler thought he was under Holden's orders to determine time and nothing else. Fraser, feeling that Keeler should be entirely under Floyd's orders, wrote, "There can only be one captain which is you otherwise there will be a conflict, which would be bad for the observatory."

Luckily Fraser's fears proved unfounded. Keeler stayed on. Floyd found him "*mighty* well instructed, sound, ingenious and enthusiastic."[57] An amiable man of few words, he was a good-hearted, cheerful fellow. Brashear, who knew him well, commented that he could do the eating and drinking for a pretty good crowd. He was a man after Floyd's own heart, and the two men soon became close friends.

Keeler, fourteen years younger than Floyd, felt in a way like his son. Floyd gave him in 1887 a folder for his (scientific) correspondence. On the inside front cover he had painted his steel cutter off Kono Tayee with "Single Reef, and 2nd jib." In the background loomed Mount Konocti. Underneath he wrote a verse:

> When out upon the lowly deep
> We think of all around us,–
> Let's also in remembrance keep
> The star *within,* that binds us.[58]

Born in La Salle, Illinois, Keeler had moved with his family to Florida when he was twelve. His father, William Keeler, had sailed around the Horn to California during the gold rush, then on to China and around the world.[59] In 1861, at the beginning of the Civil War, he became a paymaster, first on the famous *Monitor,* then on the USS *Florida.* Aboard that ship he must have been keenly aware of the activities of the Confederate sea raider CSS *Florida,* on which Floyd was serving. At the war's end the Keelers moved to Mayport, a town at the mouth of the St. John's River in northern Florida, on the old road to St. Augustine. Here they lived in a "sylvan home embosomed in grand forests of palmetto," with a fine ocean view. William Keeler grew pecans and oranges there and ran a small general-repair business. Young James showed an early fascination with instruments. At eighteen he surveyed the mouth of the St. John's River and built a telescope and transit for "astronomizing." In 1877 he entered Johns Hopkins University in Baltimore. He was largely self-educated, without a high school education. In the summer of 1878 he joined the eclipse expedition to Central City, Colorado. He told afterward of the hot, dusty, smoky train trip from Baltimore to Denver with a rough, whiskey-drinking crowd, of the stop in Topeka, where he "smashed" the waitress, and of his arrival in Denver, where he met Edward Holden, whom he had first seen when he had been a student at Johns Hopkins.[60]

So the work went on at a lively pace, but the long agony of building the observatory was telling on both Fraser and Floyd. In the fall of 1886 Fraser had to leave for the city. He had caught a bad cold, had an ulcerated throat, and felt much "pulled down."[61] Although he found it hard to leave, he said the trustees were lucky to have Captain Floyd to direct the work on the mountain. "The Lord only knows when I will come back!" Fortunately his fears proved exaggerated. Two weeks later he was back at work with Floyd, figuring out plans for the work still to be done.[62]

9

James Lick's last journey

On January 10, 1887, the Mt. Hamilton Stage rolled up the mountain, bearing James Lick's body to its final resting place under his great 36-inch telescope.[1] At the observatory's front entrance Captain Floyd was waiting to receive the august "Committee of Escort"–representatives of the University of California, the California Academy of Sciences, and the Society of California Pioneers, with the president of the Pioneers, the Lick trustees, and the governor of California. He led them into the handsome rotunda, and there they turned the casket over to him as president of the Lick Trust. Then, according to a San Jose reporter, Floyd asked Fraser to remove the lid so all could see Lick's remains through the glass. Fraser looked down and said softly, "That is what is left of Mr. Lick."

Voices asked, "In what way do you recognize the remains, Captain Fraser?" Fraser answered, "I recognize him by the form and cut of his whiskers, which are very bushy and thick and came up close under his chin, as he had a habit of stroking them. The only difference I can see in his whiskers is that they have grown longer and higher up on his cheeks. I also recognize him by a plain pearl button in his shirt busom."[2]

Floyd then closed the casket, sealed it, laid on it an elaborate identification book, and draped it with an American flag. When the dignitaries had gone, he locked the rotunda doors and put the keys in his pocket. Afterward, in his "log" Fraser wrote of James Lick, "A watch was set and we left him within the walls con-

James Lick's Conservatory in Golden Gate Park, San Francisco, July 4, 1888, program.

structed by his wealth and dedicated to science."[3] The next morning everyone gathered in the great dome to watch the casket as it was slowly lowered into the vault of brick and stone that was to support the iron pier for the 36-inch telescope. In that vault, Fraser noted, "his last home, Lick's mortal remains would remain for ever and ever." According to Floyd, "The Trustees have concluded to place his remains in this pier, believing that the most powerful telescope so far made in the world will make his most appropriate monument, and this commanding site overlooking his California home his most fitting resting place."[4]

On the simple bronze tablet at the base of the pier appears this inscription:

HERE LIES THE BODY
OF
JAMES LICK

In a rough draft of his speech, Floyd had written, "His true monument is the Observatory which he reared, and his lasting memorial will be the results of those astronomical observations, which his generosity has instituted and endowed."[5]

Sometime later Holden, fearing Lick's remains might interfere with the telescope foundation, wanted them removed. Fraser protested vehemently, "Let no whim of some martinet temporarily in authority deny the simple wish of the dead that his bones rest beneath the structure which his munificence reared for the welfare of the whole world."[6] Floyd was equally shocked. He recalled his last minutes with Lick when the old man had repeated over and over how pleased he was to have decided on the resting place for his body, emphasizing that the suggestion of a tomb so satisfactory had relieved his mind of a great worry. The trustees, in obeying Lick's expressed wish, regarded it as a discharge of a sacred trust. "Should his body be removed now simply to suit the ideas of some living man, it would be an outrage and a desecration." The trustees had a duty to prevent such a move at any cost.

Another view was taken by others, closer to the millionaire: "He will be buried in the base of the pier of the great equatorial on Mount Hamilton, and will have such a tomb as no old world emperor could have commanded or imagined."[7]

Still others took the millionaire less seriously. An old Californian, visiting the observatory, told the astronomer in the dome that he "knew Jim when he was in no fix to give away telescopes, nor nuthin' else." When it was suggested that once Mr. Lick did get ready to donate a telescope, he certainly gave a good one, "Our perverse and argumentative friend [remarked], 'Yes, and he lies dead at the bottom of it.'"[8]

On February 16, 1887, as twenty-two inches of snow fell, piling from three to five feet around the observatory, the Floyds arrived on the mountain. The Captain noted, "Madam and I will probably have to foot it up from the brick yard."[9] The road to the top was impassable.

On the mountain Floyd took time to write to Holden on an entirely different subject. He had long felt that the university should offer to those who wanted to enter the marine profession the same facilities it did to those who wanted to become "lawyers, divines, doctors, chemists, mining and civil engineers, agriculturists, horticulturists, and astronomers." He believed investigation would show that more than half the marine disasters occurred "from the want of education of the Commanders in the higher branches of navigation, and frequently from gross ignorance of the simple elements of natural philosophy."[10] Therefore, he urged that merchant marine officers be given the chance to gain a commission in the naval reserve after passing the examinations required by the government. To carry out this program he suggested that his old friend Captain William H. Parker be made instructor of navigation and nautical astronomy at the university. Parker, author of *Elements of Seamanship,* had taught at the Naval Academy, then was superintendent of the Confederate Naval School before becoming captain of a Pacific Mail steamer, plying between San Francisco and Panama. Floyd assured Holden that if he started this branch of instruction at the university, he would be doing a good thing that everyone would thank him for someday. Long afterward Floyd's dream would be realized when the naval unit was formed at the university.

Another important step was taken at this time. In answer to a request from Floyd and the Lick trustees, the California state legislature showed "its entire appreciation of the Observatory and its work, by passing a resolution providing for the issue of such reports, observations and researches, as may with the approval of the Governor of the State, be submitted by the Lick Trustees, of the Regents of the University, for publication." [11]

Meanwhile Fraser was pushing work on the brick wall for the dome. On June 17, 1887, Floyd noted that "the strongest brick wall I ever saw is now nearly level with the plain of the observatory. Every body is busy and rushing." But Tom caught a bad cold and was "much pulled down." He finally had to take time off to recuperate in San Francisco. Floyd carried on in his absence. Early in October a reservoir on East Peak was completed; 173 feet above the observatory floor, it held 36,750 gallons and was hooked up with a complete system of supply and discharge pipes. Floyd, thinking of his windmill at Kono Tayee, had decided to build a 16-foot windmill to pump back the water used for motor purposes. By October 8 it was working beautifully.

After the windmill was set up, Floyd wrote enthusiastically to the maker of this "Golden Gate" pattern windmill. He explained how it worked.[12] "This windmill has to perform the work of forcing water from a reservoir 31 feet below the plane of the Observatory through 3500 feet of 2½ inch pipe to a reservoir 194 feet above the same plane and which furnishes the hydraulic power for rotating the 36 inch equatorial. Against a head of 205 feet, and the friction incident to this considerable length of pipe it does work well with double acting force pump of 3 inches diameter of piston and 5½ inches stroke. Even in very moderate breezes, its elevated position of about 4200 feet above the Santa Clara valley, exposes it to the full violence of the prevailing winds which it has now done for six months."

By January 26 the girders and shutters for the dome were on the "hill." [13] Early in April Floyd exclaimed, "The Dome is coming and the hill is getting full of iron." [14] Men from the Union Iron Works were there, ready to set it up. Fraser, eager to finish the job as quickly yet as perfectly as possible, complained that these men were the slowest workers he had ever seen. He added, "If these men would have less womanizing and less Hunting and attend to their business it would get on faster."[15] Nevertheless, by June 7 the frame of the dome and the elevating floor were up and in place. The riveters were driving away. By the end of August all the staging was out of the dome. Floyd was elated. "It looks splendid," he said.[16] Soon afterward Mathews appeared on the mountain to photograph "Heaven and Earth."[17]

A reporter for the *San Jose Daily Mercury* crowed, "The California firm not only submitted by far the best design but was the lowest bidder, and the sequel shows that California is in the front rank of progress. Let all honor be given to the men who have shown to the world when occasion required any great enterprise to be done, that our state will stand in the front rank, not only in gold, grapes, wheat, barley, quicksilver, but also in mechanical pursuits."[18]

Meanwhile Floyd and the trustees were becoming concerned about the third lens or photographic corrector for the 36-inch, an idea proposed by Holden late in 1885.[19] This corrector would convert the telescope from a visual to a photographic refractor, bringing all the yellow light to which the human eye is most sensitive to an instrument corrected to bring all the blue light to a focus.

The trustees had hoped to obtain the disk for this corrector from Feil in Paris. Then Alvan G. Clark in Cambridgeport would figure it, with the final tests to be made with the corrector attached to the 36-inch refractor. But there was trouble with Feil and the crown blank he delivered to the Clarks. They felt that the internal strains would cause the blank to break. This is exactly what happened.[20] After this the trustees waited anxiously for months for better news. The situation became worse with the death of the elder Feil.[21] The firm was taken over by Mantois, but no one knew when or if he could cast the needed disk.

On March 7, 1887, Mantois cabled Clark from Paris, "Continue disque et affaires. Letter sent."[22] The Henry brothers in Paris tried to reassure the Clarks. But Floyd was profoundly troubled. He wrote Alvan Clark, "I am very much upset about this 3rd lens business and dont know what to do." [23]

The Clarks suggested that Floyd consider a 69-centimeter flint disk owned by Yale.[24] But it was marred by an excessive yellow cast, and it was expensive. Floyd did not like the idea. Still he was anxious to solve the problem and finish this final step in the building of the observatory. On June 1 he told Mastick of his despair of ever getting a large crown glass in Europe "except at enormous price and disastrous delay, unless Mantois had one nearly ready. I believe that to depend upon him would be leaning upon an uncertainty of one or two years." [25] Before making a final decision, he consulted Lick trustee George Schönewald, now the manager of the El Carmelo Hotel in Pacific Grove. Schönewald wrote to Mathews, "I am not in favor of purchasing the Yale flint glass, and furthermore I am not in favor of having anything whatever placed in the Observatory that is not strictly first class in every respect. It seems to me that the Lick Trustees should order a perfect glass to be made, and should it require from one to two years time the fault will not be ours." [26]

That settled the matter.[27] Soon Alvan G. Clark was on his way to Paris to see Mantois. By September 2, 1887, Floyd was able to tell Holden[28] that he had received a letter from Clark in Paris, written on August 13.[29] Clark reported that he had Mantois at work on a very promising block of glass he had agreed to mold for the trustees. It weighed 165 pounds, had no visible fault, and would mold 34 inches in the round; that is, it would make 32 inches clear aperture. He thought it would make a fine disk.

Just as Floyd finished reading this letter, the telephone rang. A telegram had arrived from Clark in Cambridgeport: "Glass just molded – satisfactory contract effected." [30] A suffocating weight was lifted from Floyd's shoulders.

In May trustee Plum visited the Clarks, who were figuring the photographic corrector lens. He reported enthusiastically, "I do not think that anywhere except in the stories of mythology can there be found a description of such a work shop and their three delvers into the arts and mysteries of science as are represented in these Clarks. The old Gent is as active as before and now that his mind is relieved from the great anxiety of the 36″ lenses, he seems to have renewed both vigor and youth."[31] Three months later, his greatest work completed, Alvan Clark was dead at eighty-two.[32]

Afterward, in Cleveland, Plum visited Warner and Swasey, who were pushing work on the telescope mounting. He wrote to Floyd, "Evidently there has been more thought and work required to invent and plan out the necessary machinery than W & S ever imagined and they have not spared any pains or expense to outdo any other machine of its kind."[33] One of the greatest sources of trouble was the hard steel axis. They had to make three castings before getting a perfect one. On October 12 Floyd arrived in Cleveland. There he was joined by Newcomb from Washington, Burnham from Chicago, and John Brashear, the optical expert from Pittsburgh. Holden had been invited but could not come. They spent all afternoon at the shop. Floyd noted in his diary, "All hands busy getting mounting of 36-inch telescope ready for our inspection tomorrow." Looking up in awe at the huge mounting, weighing some thirty-seven tons, soaring high above the shop floor, he exclaimed, "The tube looks like a Mariposa big tree. It's a whale."[34] Burnham told Holden, "I believe *you* will be astonished when you come to see the thing up, at its enormous size. It is almost sublime."[35]

On October 16, after two more days at the shop, Floyd reported to the trustees, "Prof. Newcomb has made a thorough inspection with me. The design is excellently executed and the work very satisfactory. Will be shipped before first November. Railroad promises 14 days to California." That night he left for Pittsburgh with Simon Newcomb. On their arrival a very old, small omnibus took them to Brashear's shop in Allegheny City. Here Brashear, now forty-six, had worked as a "greasy" millwright until the day he came across Henry Draper's famous paper on making reflecting telescopes and began grinding a mirror. Discovering the joy and sorrow of making telescope lenses and mirrors, he had begun to design astronomical instruments for others.

From the first, Floyd had been fascinated by this man with

twinkling eyes set deep under heavy eyebrows who, with his flowing black tie and loose coat, looked more like an artist than a businessman. As he came to know him better, he found that Brashear was in many ways as he appeared.[36]

Soon after their arrival the noted astronomer and director of the Allegheny Observatory, Samuel P. Langley, came in, and they spent an interesting hour or two in Brashear's shop discussing plans for the large, powerful star spectroscope for the 36-inch telescope.[37] It was designed by James Keeler and was the first of its type ever built, with a large effective aperture, making it fast and capable of giving bright images. Langley, who had long been interested in the subject of flight, then led them to the observatory and showed them his pioneer aeronautical experiments.

Floyd lunched with Brashear and his wife, Phoebe, and after another hour at the shop Langley took them for a drive, "bringing up" at the Pittsburgh Club for a charming dinner. That night, after "one of the most delightful days," Floyd left for Philadelphia and Cambridgeport to see Alvan Clark and arrange for the transportation of the third lens to California. The weather was cold and nasty as he boarded the Fall River steamer for New York. On October 24 he found himself back in Cleveland, and on October 25 he returned to the shop. The next day about five thousand people came to see the giant telescope mounting. Among them Floyd met many distinguished people, including "the very handsome and intelligent Mr. Brush, the electric light man."[38] After the long months of agony, these were deeply rewarding, intensely exciting days. Floyd then headed for California on a trip frustrated by delays caused by a serious accident involving a freight train. He arrived home just in time for a sad parting.

Tom Fraser had expected his work for the observatory to be finished by November 1, 1887. He had promised to become the business manager for Miller and Lux, the owners of a huge ranch. He was forced, therefore, to leave the mountain. "I will commence now to pack up and leave here for good."[39]

On Sunday, November 30, 1887, he wrote in his "log." "Quite breezy today with lots of peopole at the mountain summit, painters working in large Dome today."[40] He ended without salutation. But it was hard. Fraser had given his heart and soul and the best years of his life to the building of the observatory. Life elsewhere would never be the same. In one of the last entries in his "log" he had written, "We are laying walks around the summit to walk so as astronomers can keep out of the mud."[41] His heart remained on Mount Hamilton. One day he wrote wistfully from Bloomfield Farm near Gilroy, "You will I suppose often think of me as the stuff goes up. I often think of you and the folks up there, and look up to the Large Dome with a longing eye."[42]

In 1886 Holden had published a graphic and balanced account of the Observatory in the *New York Tribune.*[43] In it he described the amazing transformation of the mountaintop in the seven years since the one-room cabin had been built for Burnham, with its small dome perched insecurely on a narrow ridge between two sharp points of rock at the summit. He told of the small village that had grown on the mountain, with the little white houses and workshops for the employees "nestled together" on a narrow saddle just below Observatory Peak. "Nothing," he pointed out, "has been built for mere show, and yet no expense, however lavish, has been spared to make the whole observatory – buildings, instruments and equipment, perfect." Looking back on that building he wrote magnanimously of the debt owed to Captain Floyd for his direction of the immense undertaking, noting that Floyd was so situated that he could give his whole time to the complicated duties of the Lick Trust. (He also could have pointed out that without Cora's fortune this might have been impossible.)

"The observatory has been from the beginning under his direction, but in addition to this he has administered the whole estate of over $3,000,000. so as to bring the best results not only to the observatory but to many other institutions founded or endowed by Mr. Lick in California."

Holden praised Tom Fraser, too, for his able assistance to the Captain and told how Floyd and Fraser together studied the details of every other important observatory in the world, with the Captain inspecting many of the best observatories in Europe. Holden gave an accurate picture of the observatory plans formulated in Washington in 1880 by Floyd and Fraser, in consultation with Newcomb and himself. He told of the selection of the Mount Hamilton site. "Seven years ago there was only a trail over the mountains, which was unvisited except by hunters. Now there are strong low buildings all around full of beautiful instruments which represent the finest and most delicate mechanical conceptions. All of these have been brought from the various parts of the world, and have met and fitted together with perfect precision. The clocks from Amsterdam and London are side by side with meridian instruments from Hamburg and Washington." He envisioned the boundless possibilities for the Lick refractor, as the telescope, with the addition of the third lens, would be turned into a giant camera.[44]

At that point Floyd and Holden apparently were still friends. But gradually, like the dark shadow of an eclipse, a change came over the astronomer; his resentment and jealousy were soon to explode in a bill dated May 1, 1888, for services rendered against the trustees. This extraordinary turnabout took place as he became increasingly angry that he had not yet been allowed to take over on the mountain. He was still only the prospective director of the observatory and could not assume that post until the Lick trustees turned the completed observatory over to the regents. This was a fact he apparently refused to accept. While living in San Francisco, he tried to give the impression that he was already in charge, writing letters, publishing articles, recounting the progress of work and details of life on the mountain. As a result, Floyd found his warm feelings gradually changing, and Holden's relations with both Floyd and Fraser had deteriorated.

Before long they also learned that Holden was not proving the success as university president they had hoped. His rigid, domineering, humorless attitude and intolerance of student foibles soon got him into trouble. According to V. A. Stadtman, historian of the University of California, Holden could neither understand young men nor sympathize with them. Their harmless escapades were crimes "to one in whom the reverence for order and precision is so rooted as in this inflexible combination of martinet and astronomer."[45]

As a result of these and other troubles, the astronomer became more eager than ever to exchange his presidential domain for Mount Hamilton, which he had visited soon after his arrival. Yet, for many months, he was not allowed on the mountain. Fraser, sure his presence would interfere with their work, had written anxiously to the Captain, "I hope you will not encourage Prof. Holden to come and stop here till we leave. Let us have the place to ourselves for Gods sake keep him away as I hear by his speech in the newspapers that he is to come the 1st of July. It makes me sich to think of it."[46]

When, at this point, Holden, always full of ideas, urged a study by E. W. Hilgard of the agricultural resources on the mountain, Floyd told him firmly that he would have no quarters to spare or horse feed or any room for anybody not concerned directly with finishing the observatory. This, he said, was squarely his judgment in the matter, "and this must settle it, whether fortunate or unfortunate," although he personally was sorry to decline this kind of proposition. As to Holden's remark that "this matter ought to have been attended to 5 years ago," he concluded, "It seems to me that I could have about as appropriately built a whaling station on the rim of one of our spring tanks. Neither the Regents nor the Faculty of the U. of C. nor the press of California then condescended to believe there would be a Lick Observatory."[47]

Holden was upset because the smaller but valuable instruments on the mountain could not be used until the 36-inch was completed. He ignored Floyd's repeated reminder that the Trust Deed did not provide for any salaries except those of the Trust employees. The regents could not take on such expenses until the observatory was turned over to them.

Floyd, too, was becoming increasingly annoyed by Holden's attitude. As he had told him repeatedly, nothing could be done toward beginning regular astronomical work until the Lick trustees were through with the dust of freight and handling heavy

Sealed tomb of James Lick at base of pier of great telescope on Mount Hamilton. Building blueprint with Thomas E. Fraser. (Reproduced by kind permission of Mary Lea Shane Archives of the Lick Observatory.)

iron and wood work, clearing scaffoldings, and painting and cleaning the observatory. When he turned the observatory over to the regents, he wanted everything shipshape.

One day, more exasperated than usual, he sat down at his desk and drafted a letter to the future director. "It is quite true," he wrote, "that this Observatory is to belong to the State of California and to science but it don't yet. It will not be turned over to either until it is completed. I have made the mistake of letting this point escape my view sometimes – always with unpleasant consequences. I propose and am determined to arrange everything pleasantly for the balance of time that the Lick Observatory is under my direction."[48]

Floyd did not send this draft. But, three days later, in answer to another of Holden's "uncalled for" letters, he wrote, "It is true that you have rendered most valuable assistance to the Trustees in constructing the Observatory through your Advice as an astronomer, but so have many other astronomers of the world – and it is also true that I have selected you out of all of these to receive the splendid reward of becoming its future Director." He concluded cryptically, "I will let you know when your being here might be of service to the Observatory, and when it will be convenient to me. . . . As to personal embarrassments, I intend to prevent them by not permitting an occasion."[49]

Still, with that mixture of intensity and tenderness that characterized his nature, Floyd hated to see his feelings for Holden change. One day he had written from the Union Club in San Francisco, "It makes me sorry to see ever, shadows of trifling things upon your expression. I did now and then yesterday."[50]

10

Final stages

On a stormy day at the end of November 1887, Floyd waited impatiently at the San Jose station for Ambrose Swasey and his wife to arrive from Cleveland. It was "blowing a howler." Together they waited anxiously for several days for the "telescope cars" that had somehow disappeared along the way. Floyd wrote, "I have my shear legs up and all purchases and tackle rove off for a hoist the moment I get any of the parts on the hill." [1]

The *San Jose Daily Herald* reported, "TELESCOPE – One carload of Machinery Safely Arrives. More Will Soon Be Here. Valuable tubes, Clockwork and Mountings which Will At Once Be Placed in Position. Delay in Arrival Caused by Freight Blockade. Uneasiness Felt by Floyd and Other Trustees." [2]

One of the freight cars apparently had been blocked at Truckee. The other arrived in San Jose on December 6. Years later Swasey recalled the day they loaded the huge sections of the mounting, weighing several tons, on the trucks "drawn by six or eight unfortunate beasts, called horses." [3] On each truck there were two drivers, one holding the reins and riding, the other walking along beside the animals, pelting them with stones to revive their spirits as they climbed the many long and steep grades up the winding road. As Swasey commented, "One could easily imagine the poor creatures – nearly dead from exhaustion – saying to one another, 'This may be all right for science, but it is mighty hard on horseflesh.' " [4] The job of getting all this iron and steel up the "hill" was obviously tremendous. It

The 36-inch telescope on Mount Hamilton, contemporary etching, 1888. James E. Keeler is at the eye end of the telescope. R. S. Floyd is at the left and Ambrose Swasey at the right on the floor of the telescope's dome. (Reproduced by kind permission of the Mary Lea Shane Archives of the Lick Observatory.)

was made harder by the horrible weather that brought wind, rain, snow, and slush. Everything wore "a most gloomy and chilling aspect."[5] For two days they could not even see the nearest reservoir. Time lost all meaning – Floyd begged Mathews for a small calendar. "I have nothing to tell the day of the month."[6]

From the top of Mount Hamilton Floyd, who had gone ahead, watched eagerly with his spyglass, as the huge sections of the pier moved through Hall's Valley, then slowly wound up the mountain. "It is beautiful," he told Mathews, adding, "I go to S.J. Tuesday to meet Clark and bring up the 3rd lens."[7] Afterward he wrote to Newcomb of this final stage in their long adventure, "I wish mightily you could see this Observatory before I get through with it."[8]

On December 13 the two heavy sections of the mounting finally arrived after a hard journey. Said Floyd, "I have successfully mounted them both and have this moment landed the head in place. It weighs 8,112 lbs. The heaviest piece remaining to be mounted weighs 5000 lbs. . . . The big pier now being complete looks fine."[9]

He had taken no chances, making the "rigging" for the mounting five times stronger than he felt absolutely necessary. If any of those 3- to 4½-ton weights should, from an unseen flaw in a hook, or careless lashing, drop 37 feet onto the elevating floor, the consequences would be disastrous. It would mean a loss of $50,000 and a year's delay.

At last, after years of uncertainty, everything was on the mountain. "I hope," said Floyd, "to launch this Observatory into the ocean of science about in February."[10] At noon on December 24, after working from 6:00 A.M. until bedtime without a moment to spare, he told Mathews that at that moment they had landed aloft and safely in place the last section to complete the big tube. "It looks very majestic!"[11]

They had hoped to have the tube ready to receive the big lens by Christmas Eve. But the "emaciated" beasts of Wandell's teams were slow, and some of the material had not yet arrived. Floyd was "in a fury."[12] While he waited, a letter brought the bad news that his house in San Francisco had nearly burned down some days before.

Meanwhile, on the mountain he was checking on the center of motion of the telescope with respect to the dome and movable floor. To his surprise he found that the masonry foundation was about a foot too high above the movable floor. Yet he saw that, even at the zenith, the edge of the cell containing the photographic corrector lens would clear the girders of the dome skeleton by nine inches. He scribbled his observations on the back of an envelope that already contained notes on the terms of the Warner and Swasey contract. Perhaps he had been thinking of the firm's responsibilities according to the contract.[13]

A telegram then came from Alvan G. Clark, the son of the elder Alvan: "Glass wife and myself on the way." On December 29 they arrived on the mountain through snow and freezing cold in this, one of the stormiest winters on record.[14]

The final stages of work on the telescope were dramatic, and the drama was heightened by the violence of the elements. On New Year's Eve Floyd, with Swasey, Keeler, Clark, three workmen, and the watchman, hoisted the 36-inch objective onto its carriage and wheeled it into the dome. They soon discovered that the snow and the bitter cold of the storm that had raged since December 26 had stuck both dome and shutter. With Swasey's and Keeler's help Floyd was able to move them a little before dark. At 11 P.M. a few stars shone out at the zenith, but Floyd decided it was too "moist" to try the telescope. At midnight he gave up "and with patience turned in at one past midnight having seen the old year out and the New Year in." [15] For Dick Floyd such a New Year's celebration was indeed unusual.

On Sunday, January 1, there was a slight misty drizzle; all that night it rained. The next day brought more rain, with the wind blowing from the southeast; it became stormy, with a rattling seventy-five-mile-an-hour gale blowing at dark. Floyd, who had known many a wild blow aboard ship, said this was the hardest one he had ever experienced. It was the worst weather ever known on Mount Hamilton. His visitors were terrified, with reason. Swasey found it impossible to walk from one building to another without holding on to something for fear of being blown off the mountain into the canyon nearly a thousand feet below.

At one point the wind reached eighty miles an hour. The anemometer blew off and landed in the valley below. No one knew what velocity the wind then reached.[16]

Finally, on January 3, the weather cleared. The thermometer fell to six degrees Fahrenheit, the coldest by seven degrees ever known there. Everyone hoped to "get a peep" at the stars through the telescope that night. At 7:30 they prepared for the first look, only to find the dome frozen hard in the liquid seal created by water that had been beaten into it. Luckily Floyd had turned the shutter to the southeast during the day before the dome got stuck. With a "good deal of manipulation," they caught several faint stars before pointing the telescope at Aldebaran. "He appeared like a blazing red sun." [17] Clark, Keeler, and Swasey took turns observing it, but they soon realized something was desperately wrong. To their horror they found the telescope tube was six inches too long.

The focal length of the objective lens proved to be not 56.5 feet as Clark had stated, but 56.0 feet. Therefore, the telescope tube, made to fit this dimension, was too long. For several days nothing could be done while Swasey cut off a 6-inch section from the tube and remounted the lens. Floyd, furious, dashed off "Some Notes on the height of the Center of Motion of the Great Lick Telescope and on the true length of the 36-inch objective glass found by looking at the first star January 2nd, 1888, showing an *inexcusable* mistake by Warner and Swasey in making the pier 13½ inches too high and an *inexcusable* mistake by Alvan Clark and Sons in giving the focal length of the object glass 6 inches too long." [18]

They cut off two inches from the adapter of the lowest-power eyepiece and shoved it in just far enough to get a fairly good focus. Finally they managed to get a few glimpses of the great red star before squalls of mist passing rapidly in front of the telescope turned to dark cloud. "The weather grew so thick that the heavens entirely disappeared." [19] After closing the shutter, they waited to see if there would be any change. At midnight it began to snow. Everyone turned in for the night.

On January 4 a snowstorm raged all day. Only on January 7

could the first real observations be made. It was bitterly cold. At 9:00 P.M. Floyd scribbled a note to Mathews: "We have a beautiful night and are now observing–just had a look at Rigel–Splendid!"[20] Afterward, although they could glimpse only what passed in front of the shutter open to the southeast, Swasey was given the honor of being the first to observe the ringed planet Saturn. Floyd wrote enthusiastically, "We had a magnificent look at Saturn last night. The definition was exquisite and it had the silvery brightness of the moon. All hands were delighted and turned in at 2 A.M."[21] The next night he noted, "We have this moment been looking at Neptune. There is no doubt that we have the most powerful optical instrument in the world." But the dome was still stuck, and the weather remained frigid, with the thermometer registering sixteen degrees Fahrenheit. "Everybody shivering over sickly fires of wet wood."[22]

"The big telescope works well," he added. "I can safely say excellently; but there is a world of nice adjustments to be made before we get it in the shape that will do it justice and with which such a monster every one of these means lots of climbing to great and giddy heights and great precautions to save one's neck."

To this Keeler, writing of Saturn, would add, "There was not one scientist or mechanic, of the little group of men gathered around the end of the telescope, who did not utter an exclamation of wonder as the flood of light from the glorious object in the instrument entered his eye. It was beyond doubt the greatest telescope spectacle ever beheld by man. The giant planet, with a splendor and distinctness of detail never before equalled."[23]

These early results were promising but the bitter weather brought other problems to the pioneers on the mountain. Among these were the chimneys. One day Floyd wrote in desperation to Mathews on his "fancy chimneypot" that had "gone back on us in the worst kind of a way" when the wind was due north and the thermometer registered six degrees. The chimney, he said, smoked worse than ever. "In fact it beats any thing smoking I ever did know." It not only smoked him out of his office, but it filled the whole observatory with smoke. The rough uncomfortable time continued as the "confounded frost" brought myriad things to look after. "All hands now doctoring frozen pipes in dwelling house and nobody is warm."[24]

Still another problem had arisen. Floyd demanded of Mathews, "What kind of coal oil have you sent me? It is the 'Star Light' brand, but is that the very best and safest kind of coal oil in the California market that I particularly requested you to get on a/c of danger to the Observatory? Since using the oil you have sent, we have come nearly having two lamp explosions. One last night in a raging storm which would have certainly burnt up the dwelling house if it had not been for quick and energetic action which saved a burning up by perhaps five minutes. A mere accident discovered the threatened catastrophe and holocaust in time to prevent dire calamity. I am not quite sure whether it was the fault of the lamp, a servant's stupidity, or the oil. But I am very uneasy about the oil, because several of our lamps which never troubled before, have been guilty of queer tricks since using the oil you have sent me. It seems to generate much heat, and to make gas at very low temperature." Asking Mathews to investigate the man he bought the oil from, he concluded, "None but the very *best and safest* brand of coal oil should ever on any account be permitted at the Observatory."[25]

In the midst of his troubles, Plum wrote to Floyd on the last day of 1887, offering to help him in any way he could, "I can relieve you of part of the heavy burthen, that I feel sure unaided and alone as you are at present, is very heavy."[26] He hoped that Floyd might long enjoy the fruits of "his great labour on Mount Hamilton." A letter came also from Fraser: "I am glad to hear that you have got the material on the hill for to complete the great telescope. I would like to go up very much. I have so many things to do here, but none of them amounts to a hill of beans. I see off in the distant north every day that it is clear, the white dome of the Observatory. I long very much to see you all and feel awful lonesome without you."[27] Floyd felt lonesome too, and missed his old friend sorely.

When Fraser read in the papers that the great glass was in its last home he sent his congratulations and love to Floyd and the telescope, and wished a happy New Year to all. He did not know what kind of a New Year it had been! On January 22 he came up for a visit. It was dark and rainy. The next night it rained again, but around midnight it cleared a bit. Floyd went out to the 36-inch with Fraser and Clark, but after an hour they gave up. The skies were completely covered. On January 24, in the forenoon, Floyd and Fraser observed Venus. They thought they saw markings through the atmosphere resembling continents and water. That night they observed with the 36-inch. But at 10:00 A.M. the next day Fraser had to leave.[28]

A few days later Trustee Plum arrived on the mountain. He "found the Captain at work directing those around him and hastening the completion of the telescope." He wrote "I was glad to see that he was able to continue his work so vigorously, and yet there is evidence of the strain and care which he has endured for so long a period." [29]

By February 18 the mounting was working "very nicely." The big driving clock for the telescope was also working well. With the Clark photographic corrector Floyd was able to photograph Rigel near the focus with less than a second's exposure. Everything, at long last, looked promising. Despite the fearful weather he was optimistic. He felt he could see his way clear of almost all the difficulties. All he asked was good weather. He even thought the ill winds might have blown more good than bad. He was convinced that if he had transferred the observatory to the regents in midwinter, its scientific start would have been handicapped.[30]

Meanwhile Floyd had sent the team to San Jose for the large spectroscope that had now been completed in Brashear's shop. When it arrived safely on the mountain, Floyd called it "a very beautiful instrument," although George Comstock, who had seen it at Allegheny, had thought it a very large piece to hang on the end of a telescope. On March 1 Keeler had a chance to try it on the telescope. Although everything appeared to be right, he felt that work with it, like most work with the big telescope, would be slow "simply on account of the size of all the parts." [31] Adjusting screws, which in a small instrument could be reached with the hand, were here so far apart that a stepladder had to be pushed around from one to another.

In the *Sidereal Messenger* Keeler published a luminous account of the observations of the magnificent ringed planet. "When Saturn passed across the slit and entered the field of the eyepiece he presented probably the most glorious spectacle ever beheld. Not only was he shining with the brilliancy due to the great size of the objective, but the minutest details of his surface were visible with wonderful distinctness." [32]

"Most of these," he noted, "I had repeatedly seen before with smaller instruments, but merely seeing an object, when every nerve is strained, and even then with half a doubt as to its reality, is very different from seeing the same object glowing with abundance of light and visible at the first glance." [33] The gauze ring was conspicuous, and he even discovered a new, very fine division in Saturn's outer ring with a dark shading extending inward toward the great black division – the first such discovery made with the new telescope.

The great dome still could not be turned, as the noncongealing solution that would fill its "liquid seal" had not yet arrived, and everything continued to be frozen solid. They could only observe those objects that crossed the slit. One of these was the Orion nebula. Keeler gazed at the familiar object with delight. In the widest field that could be used, only the central part of the nebula could be seen. The great outlying streamers had to be examined separately by moving the telescope. In this way he could follow them out into space "for millions of miles, until they finally faded from view." Afterward he wrote, "It shone with wonderful brilliancy, and exhibited a wealth of intricate detail which a year would not suffice to record." For the first time, perhaps, he realized fully the great light-gathering power of the 36-inch object glass.[34]

When Keeler compared these observations made with the 36-inch with those he had made on Mount Hamilton with the 6½-inch and the 12-inch equatorial, he found that objects entirely beyond the reach of the 6½-inch and the details almost beyond perception with the 12-inch were visible at a glance with the 36-inch. So they entered a truly miraculous world. With the 12-inch Keeler had found it hard to see Neptune's satellite. But with the 36-inch on the night of January 9, it was very conspicuous. He concluded, "The great telescope is equal in defining power to the smaller ones, and has in addition the immense advantage of greater light-gathering power, due to its superior aperture." In vivid terms he described the difference in working with this giant: "The observer at the eye end of the telescope, surrounded and almost shut in by a multitude of wheels, levers and clamps, finds it hard to realize the true nature of the instrument before and around him. It seems more like the cab of a locomotive than a part of the telescope. The great steel tube, as large as a steam boiler, stretching far away into the darkness, is hidden from view of its own end and the surrounding maze of machinery, and the real form of the instrument is not apparent.[35]

"There is, however, no necessary complication of mechanical parts. Each wheel or clamp has its particular purpose. One moves the telescope in the direction of the meridian, another in that of a parallel of declination, another fixes the telescope firmly in its position at any moment, and so on with the rest. Thus the astronomer can control the motions of his apparently unwieldly instrument with a facility which is wonderful, considering the immense mass to be moved. The same motions can be given to the telescope by an assistant stationed on a balcony which surrounds the top of the pier. By means of dials and graduated circles he can point the telescope at any star required without looking away from the balcony."[36]

Before leaving for the East, Swasey invited Holden to visit the observatory. He later wrote, "I shall never forget the look of wonder and surprise as he entered the dome and saw before him for the first time the great instrument. The mechanism of the finder not having been adjusted, I stepped upon the tube of the telescope to remove the finder cap, Professor Holden sang out to me, 'Is that what you do, walk on a telescope that way? 'Yes,' I said, 'I have seen a dozen men on the tube at one time.' "[37]

Swasey ended with the following statement: "He [Holden] did not realize that the great instrument before him, of which he was about to take charge, had not only an objective with the most powerful light gatherer ever made, but that from the engineering standpoint it was the most powerful light instrument ever constructed."[38]

After this Holden left the mountain, still unhappy that he could not yet take over as director. No doubt he was still jealous of the Captain, and he now found in Alvan G. Clark an instrument for whetting his jealousy. Whereas Swasey had accepted the adverse weather conditions and other difficulties, Clark complained bitterly. Clark, working in the visitor's room at the observatory, was trying to polish the photographic corrector after mounting his iron polishing tool on an oil barrel filled with bricks and sand. In that room Floyd had put mats on the floor and a small stove, but in the freezing cold it was impossible to keep warm. Clark claimed that the mountain residents, blockaded by snow, with telephone and telegraph wires down, were cut off from communication with the outside world for days on end. He claimed, too, that the only decent thing about the telescope was the object glass. The dome was worthless, the shutter the same. He had little use for the telescope mounting.

Unfortunately Clark's complaints reached the press and stirred new waves of criticism. The *San Francisco Chronicle* two years earlier had called the telescope "a worthless piece of junk machinery and a great waste of James Lick's money." According to Floyd, the *Chronicle,* which "would attempt the murder of any man's reputation for the sake of one sensational heading," had been quick to pick up the smell of scandal. It ran columns attacking the Lick Observatory and its shortcomings, blaming them largely on Floyd. The columns, copied by other newspapers, flashed across the country.

Floyd, profoundly upset by this latest onslaught, wrote anxiously to Newcomb of the "cowardly attacks." "You were very right," he said, "in your answer that I would do exactly what the Trust Deed prescribed." He added wearily, "I am tired and heartily disgusted with the contemptible worries wholly unexpected that beset the closing up of my work here. And shall be sincerely glad when the Regents relieve me of responsibility."[39] Not only tired but suffering from heart trouble as well, he had been told by his doctors that he could never get better. The objective for his 5-inch refractor, made by the elder Alvan Clark, had arrived, and he longed for the day he could retire to the little observatory he had built for it in the haven of Kono Tayee.

To counteract Clark's horrendous tales, Floyd outlined for Newcomb the facts: "Saving the weather, Mr. Clark never in his life had such splendid facilities and assistance for doing any work he ever undertook in his life. He got satisfactory photographs on every occasion but two when he tried it with the S.E. gale blowing squarely into the shutter against the object glass – circumstances under which no one in the world would ever make observations, nor would he, except that he was in a hurry to get away – and that much of his delay was his own error on the focal length. . . . Mr. Clark owes more to the Lick Trustees for patronage and courtesies, and more of his reputation, perhaps, than to anyone else in the world."[40]

One evening in February an amateur, Charles Burckhalter, director of the Chabot Observatory in Oakland (where he had a $10\frac{1}{2}$-inch refractor), threw a banquet in Clark's honor. He invited both professional and amateur astronomers, including Edward Holden. Burckhalter, upset that his name had not been included in the mailing list for volume one of the publications of the Lick Observatory, was receptive to the complaints of Clark's wife, who had been extremely unhappy during her days there. Soon Clark's complaints appeared in the press and were copied in newspapers all over the country. Many people believed that Burckhalter was not the chief instigator of this campaign. In the East Brashear thought George Saegmüller of the firm of Fauth in Washington might be responsible. But, in the West, many were sure that Holden himself, consumed with overpowering jealousy, was really the guilty one.

Keeler, appalled by Clark's attitude, called him "a terrible old Blow."[41] Like Floyd, he was sure Clark was not the only source of the newspaper storm. They both felt Holden must be the "sneak dog" instigator. At the end of February their suspicions were confirmed. Floyd's friend, James T. Boyd, lawyer for the trust, met George Davidson on the street in San Francisco.[42] Davidson asked if Floyd "had yet found out Holden." Boyd said he had traced the *Chronicle* articles directly to Holden and his friends. He told Floyd, "After all, when your work is done, it will speak for itself, and disparaging remarks, engendered by envy or malice will be forgotten, however they may annoy at first. The man who expects grateful appreciation from the public, while working for them, is doomed to disappointment."

Another friend, Valentine Gadesden, equally upset, wrote angrily, "I am mad with indignation at the articles in the Chronicle for which I have not the slightest doubt that Holden is the inspiring madman. . . . Everybody I have met believes him to be at the bottom of it." He offered to organize an attack in the *Examiner* that would make Holden see stars without a telescope. He exploded, "What a rage Mrs. Floyd must be in and Miss Matthews against Holden!" [43]

As the attacks on Floyd increased, Lick trustee George Schönewald did everything in his power to counter them through influential friends in Monterey and San Francisco. Trustee Mastick demanded strong action, proposing a commission of astronomers to investigate the charges. Floyd, hoping still to turn the observatory over to the regents peacefully, feared this would only cause delay and added expense, "and of course irritation to Mr. Holden, and would to some extent place him in a humiliating position before his brothers of profession. If he and the Regents are satisfied with what we turn over, that would bar future criticism of our work, but we might be unjustly abused over the shoulders of the commission." He thought it better to wait until the trustees received the report Holden was planning, "when it might be necessary to decide who was right." [44] With his loyalty to his friends, Floyd found it hard to understand how Holden could now turn so completely against him.

In a letter to Newcomb, Floyd wrote of the scurrilous newspaper reports, one of which even claimed that he kept a mistress on Mount Hamilton. How could he keep a mistress on that secluded mountain, where he was living with his family? If he could look at such rumors with an amused eye, he would see that they had nothing to do with his accomplishments, of which he wrote with justifiable pride and no regret, "I ventured the experiment of interesting Warner and Swasey with the largest mounting in the world and the Union Iron Works with producing the best Dome, Shutters etc." Neither of these firms had ever taken on such work before. In both cases many improvements in machinery and instruments, never thought of before, had been made. He was sure their great dome and its contents would prove that fine mechanical work could be obtained in America, both for heavy matter and the most delicate requirements of astronomy "superior to anything that can be gotten in Europe and that with a heap less nonsense. . . . The Observatory will speak for itself bye and bye."[45] Newcomb, who at first had been dubious of Floyd's ability to carry out this project, now acknowledged his pioneer accomplishment that would pave the way for the great observatories of the future.

Newcomb's feelings were echoed by Keeler. Knowing the true facts from the inside, Keeler leapt to the Captain's defense. In the *Daily Alta California* he published a detailed account of the building of the observatory. He included an effective answer to the condemnation of Floyd's "junketing trip" to Europe, which actually had been made at the Floyds' own expense. He pointed out that the buildings and instruments of the Lick Observatory formed "the most perfect appliances yet devised for the observation of the heavenly bodies – the embodiment of the combined wisdom and experience of the most eminent astronomers in the world."[46]

In answer to criticism in the press, Keeler gave a vivid account of the construction of the iron pier and of its stability, which had provided the exquisite views of Saturn and the great Orion nebula on the few nights when it had been tried. He told of the telescope tube, the movable floor, and the great dome that Swasey called "most admirably constructed." He spoke of the inevitable delays: "Only those who are thoroughly acquainted with the observatory can appreciate the magnitude of the work which has been accomplished."[47] He believed that every delay had been made with good cause and to the final advantage of the institution.

Keeler recalled reading an article on the Brooklyn Bridge. When that great engineering feat was in progress, it was stated that the structure would never support its own weight. The bridge, dedicated in 1883, was still standing in 1888. The value of such criticism was apparent. "Whether the trustees of the Lick Observatory have builded as well may be left for the decision of the years to come, but in the meantime no advantage is gained by substituting the opinion of uninformed persons for that of the highest acknowledged authorities."[48] Floyd's friends were gratified by Keeler's strong statements. One wrote to Cora, "The article will help to clear the mental atmosphere of the gossiping world always keen to suspect mystery and deep design on the part of every one holding a prominent and particularly a public position."

From Pittsburgh, Brashear, equally incensed by the attacks, wrote with equal force an article with the headline "A GREAT TELESCOPE–IT'S ACCESSORIES HAVE BEEN SADLY MISREPRESENTED. Professor Brashear Considers the Lick Observatory as Nearly Perfect as It Could Be Made." He conceded there might be faults and errors: "no wonder in so great an undertaking, but what man will condemn errors of judgment when every care has been taken to make this observatory and this telescope the greatest in the world."[49]

Warner, worried about his firm's reputation, wrote to Holden, "I cannot imagine who would be quite mean enough or narrow enough to instigate such reports, but we know there are men capable of doing almost anything in that line."[50]

At this point Warner blamed all the trouble on Alvan G. Clark. Over the years the relationship between the two firms had deteriorated, in fact ever since Warner and Swasey had begun to build astronomical telescopes and had suggested a cooperative effort between the Clarks as opticians and Warner and Swasey as engineers. No answer had ever come to this proposal. Warner and Swasey therefore had turned to Brashear for optical equipment such as spectroscopes.

When the Lick trustees installed a Clark lens in a mounting

built by another firm, Clark became intensely jealous. His jealousy was aggravated by his recent experience on Mount Hamilton.

At this point, however, Floyd and Warner and Swasey had to face the fact that there were indeed real problems with the iron column on which the telescope rested. They found that vibration affected the entire instrument. The problems were discovered when high-power eyepieces were used and long-exposure photographs were taken. They lay in the iron column, the steel tube, and the finder telescope mount attached to the tube. It was feared that the wobbling might make it unusable for research. In the small 6-inch portable telescope, the prototype of the great telescope, the iron column of the pier was hollow. On a small scale the vibrations were negligible. In the 36-inch the effect was intensely disturbing.[51]

Actually Floyd had long been worried about the resistance of the hollow iron column to temperature changes, especially when the sun's rays shone directly on it or when the shutter was opened during the day. He thought that uneven heating of the column might produce warping and had suggested that the sections of the column be cast or fitted with metal furring strips to which a wooden outer jacket might later be affixed. He proposed that an air space of from two to three inches between this wooden covering and the iron column be created to act as a thermal insulator. But two years had passed before Floyd's proposed wooden sheathing was finally installed late in 1888.

For some time, too, Floyd had been worried about the way the huge iron column rested on its foundation. He questioned its stability. It did not rest securely on the masonry but instead on adjustable iron bolts. The partners insisted that the bolts had not been tightened enough and that this was the cause of the trouble. Floyd continued to argue that the iron bolts as they were installed caused the vibrations even when the column or tube was bumped only slightly. He felt strongly that the hollow iron column should be filled with sand or brick.[52] Only after months of discussion did Warner and Swasey accede to Floyd's idea.

For visual observations, for which it had originally been

planned, the mounting proved excellent. Photographically it was a different story. It was impossible to record sharp, round star images. Late in 1888 Floyd noted, "With the Lick telescope the photographic is incidental. Had it been different we would have constructed a different mounting. If it can be gotten steady enough to photograph the planets that will do. If not, less will do."[53] Astronomical photography in those days was still in its infancy.

It was a gigantic leap for the Lick astronomers from the 12-inch to the 36-inch. As Holden pointed out soon after he took over as director, "Even the motions of the tube, as the instrument was swung about to bring it into position for observing, frightened those who first used it – with a new sense of scale and magnitude."[54]

In these dark days, as the attacks continued, Floyd was grateful for the concern and loyalty of his friends. He found the attacks especially hard to bear as his heart trouble had grown more serious, and at times he no longer felt like fighting back. Even though he was sure of the end result of their years of strenuous work, it was not easy to be philosophical.

In San Francisco the young astronomer Edward Emerson Barnard was waiting to join the staff as soon as the observatory was finished. One day Floyd went to see him in his room at 1500 Taylor Street. He was charmed by the beaming face and radiant personality of this young man who had struggled so hard to become an astronomer. As a poor boy, Barnard had spent two months in school, then had gone, at the age of seven or eight, to work as a photographer's assistant. He had made "solar prints" as he sat on a roof, turning a heliostat by hand to keep the sun's image, reflected by the mirror, on a camera lens. Fascinated by the stars, he bought a small telescope with his meager savings and obtained a copy of *Dick's Astronomy.* When afterward he discovered a comet he received a prize offered by a vendor of patent medicines. Even though he had no formal education he decided to enter Vanderbilt University. Working by day, studying and observing by night, discovering additional comets, and sleeping only rarely, he persisted until he graduated in 1887. Inspired by his love of the skies, he had, like Burnham and Brashear, started as an amateur, but soon, as a keen observer, he became recognized as a professional astronomer.[55]

Now, while waiting impatiently to go to work on Mount Hamilton, he read in the newspapers the "unjust attacks" and told Floyd how sorry he was to see them.[56] Soon the Captain made special arrangements for Barnard to make an inventory at the observatory, offering him $75 per month, with board and lodging, the same amount Keeler was getting. Barnard arrived on the mountain in March 1888. Brashear had sent his congratulations. "I trust that in going to Lick your work will be untramelled, that a bright future is before you. You will find in my friend Keeler a glorious fellow. And Capt. Floyd is a royal fellow and of course you will have Burnham and others of the same ilk that will make your mountain home a paradise. Success to you and yours all the way through." [57] Although Barnard found the observing conditions ideal, the mountain was not to prove the paradise Brashear hoped it would be.

"This magnificent instrument," Barnard wrote, "complete at last in all its details, not only stands as a monument to the man whose money called it into being, but it will remain a still greater monument to the genius of the men whose brains and energy called forth from the sand and the mines the subtle materials and formed them into this noble telescope."

But in his diary Holden claimed that "poor Barnard" had received no salary since September 1887. Contrary to fact, he claimed that the president of the Lick trustees had refused to let Barnard come to the "L.O." because he did not want to be pestered.[58] About the same time Holden wrote angrily to Burnham, who was soon to join the Lick staff, to tell him of the horrible time Clark had on the mountain, of the "unheard of" obstacles faced by Swasey, and of the elevating floor, which he called "a complete and total fizzle." [59]

Without mentioning the extraordinary weather conditions, Holden found that "fortunately the whole waterworks arrangement which Floyd had devised has completely broken down. All the pipes are burst and the cylinder[s] of the water engine are split and the piston rods etc. etc. are bent into cork screws. It is an extremely fortunate thing for now he has gone to work to take out this blundering machine and to put back the hydraulic pistons which we at first devised. The only thing they will cost is time, patience and money. There is enough money in the original fund to do this and a little more. The time, I suppose they think, lasts from now to eternity and my patience is entirely gone and I believe Floyd has lost some of his."[60]

Soon after this, Holden finished his *Hand-Book of the Lick Observatory*. Here again his own views and attitudes are strongly reflected in his portrayal of himself as the true builder of the observatory, the initiator of the original plans. Years later he would write to Phoebe Apperson Hearst, a university regent, "The Lick Observatory cost $600,000. If the trustees had appointed me Director in 1880 I could have built it for not over $400,000, and it would have been completed in 2 or at most 3 years, and moreover, it would have been *right* all through, whereas now, we shall be twenty years in patching up certain small but annoying mistakes."[61] He did not mention that according to the terms of Lick's will, it would have been impossible for any director to take over the observatory in 1880.

Now, writing of Floyd and his illness, his suffering from asthma and heart disease, Holden claimed that all this stemmed from dissipated habits and from worry over plans for the observatory that "he is not intelligent and active enough to carry out." For the last year he said he had given no advice and Floyd had taken none and "every blunder that has been made on the mountain has occurred during the past year. He got rid of my advice when he had all the specifications etc. written out–and thought he needed me no more and of Fraser in Nov./87 when he thought he could dispense with him. This was all for the purpose of being alone at the finish and getting all the credit for the whole undertaking."[62] Floyd knew Holden was angry, but he had no idea his jealousy was so intense.[63]

Trustee Charles Plum was more realistic and sympathetic: "Since 1880 the Captain had been almost constantly at the ob-

Group on Mount Hamilton, 1886. Standing, *left to right:* Captain R. S. Floyd, Allen L. Colton, Thomas E. Fraser, James E. Keeler, George E. Comstock, Cora Matthews, and Harry Floyd. Seated: Charles Plum, Cora Lyons Floyd, Edwin B. Mastick, Henry E. Mathews, Edward S. Holden. (Reproduced by kind permission of the Mary Lea Shane Archives of the Lick Observatory.)

servatory and his illness was the result of his too close attention to the work. Captain Floyd was a fine type of manhood until he was attacked by the heart trouble." [64]

One day, with the end of the Lick Trust in sight, Floyd and Fraser had taken time to go to Santa Catalina Island on one of their last missions in connection with the sale of all the Lick properties. A year later, over ten years after Lick's death, the island went for $200,000 to George Shatto. But before long it fell back into the hands of the Lick trustees. At a sheriff's sale William Banning later acquired the controlling interest in the island from the trustees.[65]

Several more years would pass before the Lick estate could finally be settled. The only trustees then living were Edwin B. Mastick, Charles Plum, and George Schönewald. In June 1895 they asked the court to be discharged of their long responsibility. A reporter noted that Captain Floyd and William Sherman had laid aside their responsibility long since "in obedience to the decree of a much higher court." He then listed the monuments reared by those faithful Lick trustees. The first of these was the statue of Francis Scott Key, author of the "Star Spangled Banner," erected in Golden Gate Park at a cost of $60,000 and the eleven other philanthropic projects "attracting the attention of the entire civilized world," with the Lick Observatory at a cost of $700,000 at the top, followed by:

Protestant Orphan Asylum, San Francisco, $25,000
Protestant Orphan Asylum, San Jose, $25,000
Ladies' Protection and Relief Society of San Francisco, $25,000
Mechanics' Institute of San Francisco, for the purchase of scientific and mechanical works, $10,000
Society for the Prevention of Cruelty to Animals of San Francisco, $10,000
Family monument at Fredericksburg, Pa., $20,000
Old Ladies Home of San Francisco, with separate trustees to manage it, $100,000
Free Baths on Tenth Street, under separate trustees, $100,000
School of Mechanical Arts, under separate trustees, $510,000
Monument in City Hall Park, $100,000

The work had gone on for twenty years. One of the last acts of the trustees was the collection of a $181,000 mortgage from the Fair estate. James Fair had paid $2,000,000 for the Lick House, built on a piece of land for which James Lick had originally paid $81. The residuary legatees, the California Academy of Sciences and the Society of California Pioneers, expected finally to receive about $600,000 but had also to assume the mortgage on Catalina Island and several other small mortgages.

Afterward a writer for a Santa Catalina paper pointed out, "It is but little known that a large portion of the present-day knowledge of the stars, the sun, the moon and the millions of other heavenly bodies is directly connected with one of the outstanding real estate sales of this century."

11

To the stars

On April 17, 1888, thirteen years after the formation of the Lick Trust, President Richard S. Floyd announced that the observatory was ready to be turned over to the regents of the University of California.[1] The Captain's long job was finished. Four days later the regents arrived on Mount Hamilton to examine *their* observatory. Floyd was too ill to be there, but James Keeler told him of their visit.[2] As they filed up the observatory steps the regents were, he said, in very good humor, surprised and pleased by the completeness of everything on the mountain. He told of Judge John Hager and of Andrew Hallidie, father of the San Francisco cable car. He portrayed T. Guy Phelps, chairman of the observatory committee, as a "dry, rough cut old fellow," spitting on the marble floor with a sangfroid that would have "sent chills down Floyd's spinal column." Keeler himself concluded sadly that he felt one period in his life on the mountain, perhaps the pleasantest, was past, and he was bracing himself for what was coming. He added, "It seemed as if I must be mistaken in some way, and that this was not the L. O. Even when my thoughts were occupied I unconsciously knew that there was a vacuum somewhere which could not be filled." [3]

On April 30 Charles Plum wrote to Henry Mathews, "The more I investigate the greater my wonder at such a transformation from a scragy mountain peak to the greatest institution of science on earth and the diligence and patience required to

The Lick Observatory, on Mount Hamilton. The 36-inch refractor is in the dome at the left, the 12-inch in the smaller dome on the right. (Reproduced by kind permission of the Mary Lea Shane Archives of the Lick Observatory.)

accomplish so valuable an adjunct to the development of civilization is justly due to Capt. Floyd and his noble assistants."[4]

Ambrose Swasey, who had come from the East for the ceremony, called the day the lonesomest he had ever spent in California, owing to the Captain's absence, feeling as he did that in some way the Captain was part of the mountain. "You can't imagine," he wrote wholeheartedly, "how much we missed you."[5] Edwin Mastick said it seemed he had lost "some friend or relative." He felt like a burglar, going around and unlocking places when "The Captain" was not there.[6]

Two weeks after turning over the observatory to the regents, the Lick trustees were astonished to receive from Holden a bill for $6,000 "for expert services rendered since 1876." He wanted to set the record straight by specifying those services. "This I owe to Astronomy and Astronomers – it should not be falsely believed that the L. O. was made from nothing by Floyd alone – not even by the active and devoted (tho' often ignorant) zeal of Fraser." His bill ran to nine legal-size pages. His claims ranged from *his* plans for the observatory to those for the telescope mounting and dome, in which he said he had taken a leading part. It included a request for recompense for letters written on behalf of the trustees; for advice given to Floyd and Fraser, as the "main scientific adviser" to the trustees; for travels to observatories and instrument makers, in which only his railway and hotel expenses had been paid; and finally for his trips to Mount Hamilton and his work on the ordering of the meridian circle, the Fauth transit, and the clocks.[7]

Just a year earlier in Cleveland Simon Newcomb, Floyd's *chief adviser* from the beginning, had given him a letter marked "Confidential." It was dated Washington, April 7, 1887. At the time he urged Floyd not to make it public, but said that, as it related to business connected with the Lick Trust, he should feel bound to keep it among his records. In this letter Newcomb lists certain vital facts:

> Firstly that I was the original backer of Holden, bringing him to the notice of D. O. Mills first, then to Floyd.
>
> Secondly that in my imperfect and humble way I rendered what assistance I could in the researches and investigations necessary in procuring the objective and preparing the contract with the Clarks for the telescope.
>
> Thirdly that the original plans of the Observatory which I believe that have since been but slightly modified were drawn up in my office, under my direction and, at your request, by an architect selected by myself. The only part which Professor Holden had in the matter being that he was one of several persons kept in consultation upon the matter.[8]

Now, when Floyd learned of Holden's bill, he was deeply hurt. He considered it *rascally* and an insult to the trustees. He thought it was shaped to place on record what in effect would be a falsehood – whether allowed or disallowed – and it ought to be fought in the courts. He asked Mastick whether Holden could be dislodged, pending a court settlement, and wondered if this would delay the delivery of the Observatory. A court of law was perhaps the best and only sure way of settling anything with such a man.[9]

A few weeks later a letter arrived from the Naval Hospital in Washington, where Newcomb was suffering from an obscure form of exhaustion. "Had I not been nearly knocked off my pins myself I should long since have written to say how sorry I am to hear of your ill health." He hoped, he said, Floyd would "live long in the enjoyment of the honor due you as builder of what, I hope, will soon be one of the most renowned Scientific Institutions of the world." At the end Newcomb added, "I have this moment been astonished to hear a report to the effect that Holden has brought suit against the Lick Trust for $6,000 for services. I can hardly believe it. If he has, I suppose he will renew his claim that he planned the Observatory. Then I hope your counsel will ask him in what room or office he drew the plans." [10] It would have been impossible, he noted, to make any such plans in 1876, as Holden claimed, when the site had just been chosen and he could have had no idea of the topography of the mountain.

As Newcomb's confidential account clearly shows, Holden, despite his extensive claims, had played a secondary role in the actual building of the observatory. Yet, for the rest of his life he would continue to publish articles and books upholding his claims on the observatory planning to such an extent that some modern historians, reading these published accounts only, have taken them at face value and have given him the credit and honor he always craved.[11] Fraser, learning of Holden's latest actions, was shocked. "It is an outrage," he said, "to make a claim in that way to you who has always been his friend." [12] He suggested to Floyd that Judge Oliver Evans be engaged to defend them against Holden, "for he *knows* that it was you that managed the whole affair and everything connected with the Observatory!" "As for me," said Fraser, "I was paid for what I done, but you was not, and therefore you deserve the credit." Floyd knew they deserved equal credit. Without one or the other the observatory would never have been built.

Vehemently countering each of Holden's claims, Fraser destroyed them point by point. He exploded, "He [Holden] did it for glory and he thinks he is getting it by making a statement so long of what he done, that a stranger reading it would think the L. O. was built under his advice. It is not so, however, the Observatory was built under the management of the Lick Trustees and there is not a sub-lieutenant that can come in now and reap away glory in their own imaginations, which will hardly go down with Californians."[13]

Fraser, worried about Floyd's health, suggested a voyage at sea to take him back to the long ago time when he had trod the quarterdeck. He hoped this might restore him to his "wonted health and strength." [14] Floyd feared this would never be. His doctor had told him that, in addition to an abnormal heart, his left lung was congested, his "venous" circulation arrested. The doctor said he had seen cases end with complete recovery. Others had lived to enjoy life with some care. There was hope still.[15]

On May 15 Floyd signed at Kono Tayee the deed conveying all the property on Mount Hamilton to the university regents. It was signed on May 31 by the rest of the trustees in San Francisco. That day, Tom Fraser happened to drop by at the Lick Trust office just as the meeting was about to begin. He was amazed to find Henry Mathews and all the trustees, except Schönewald and Floyd, gathered there with regents Hallidie and Phelps, their attorney, John B. Mhoon, and James T. Boyd, attorney for the Lick Trust. It seemed an act of Providence. Fraser had no idea he would be in at the "final end" when the observatory was transferred to the regents, with the title papers and $90,000 in gold.[16]

Fraser was, no doubt, even more amazed when he was asked to deliver the papers on Mount Hamilton. In San Jose the following morning he telephoned to Holden to say he would be "on the hill" at 6:00 P.M. If agreeable, he would like to say a few words in the large dome to Holden and his "professors." [17] On June 1 Holden recorded in his diary this event he had been so eagerly awaiting. "Telegram received, Deeds delivered. Transfer consummated. You are authority. [Signed] H. E. Mathews." [18] On June 2 the new director wrote, "Delivered regulations of L. O. to

Mr. Keeler and Mr. Barnard, Fraser and Mrs. Fraser. Mr. and Mrs. Morrison leave San Jose at 1 to arrive to 6 P.M. and spent the night. Fraser is the bearer of the official papers. Fraser wanted to have the astronomers assemble in the large Dome and make a speech at delivery, but I said I would take the papers in my office, which was done." [19] Fraser, with his love of ceremony, must have been deeply disappointed.

Before going up the mountain Fraser had dashed off an article on the observatory's history and had taken it to the San Jose newspapers. He knew it contained errors, but as he told Floyd, "the main thing I went for is not far astray, that is I wanted to give Holden's professors my definite idea of who built the Observatory. I spoke of you Captain what I believe to be true and I put on record about you and what no man dare refute." [20] Afterward he hurried up the mountain. Holden met him at the main entrance, treated him well at dinner, then asked if he would still like to say something – in his office. There Fraser had the chance to tell the "professors" something about the history of the observatory – but without the final ceremony in the dome he had dreamed of.[21]

After Fraser left, Francis M. Roby, Floyd's old friend from Confederate Navy days, who had joined the Lick Trust staff in San Francisco and now superintended the work on the mountain, told Cora Floyd that he had remained on the mountain until after the transfer "and was kindly and generously asked to remain over by the hypocritical Director." [22]

Soon after his visit to the mountain, Fraser thanked Floyd for the news that he was improving. Everywhere he went, he said, people hailed him on the street, asking how the Captain was getting along. He felt sure Floyd's strong constitution would pull him through. It seemed impossible he could leave "this sphere" for years to come. He added, "Just think of it Captain I am one year younger than you and I feel today better and stronger than I ever did in my life, and you know how I came near to breaking down at the mountain only for your kindness in letting me away for a while I would ere this be across the river." [23]

On June 27, in a formal ceremony at the University of California commencement in Berkeley, the observatory was turned over to the regents. Fraser wrote from Bloomfield Farm of the account he had read in the newspapers. His wife was there, but he was not. He told Floyd, "I hear that Holden was not allowed a chance to say a word, I enclose a peace cut out of the Post which I think is very good and he deserves it all but it must make him mad to think that he is loosing his friends. That letter from Newcomb that you showed me ought to make him go insaine. it looks to me that his race is run."[24] At that ceremony Mastick read a message Floyd had written for the occasion. In a brief historical review Floyd showed how Lick's desire for the most powerful telescope in the world had been fulfilled, and he concluded dramatically, "Time in such a work is as insignificant a consideration as would be the danger that the heavens might fly away before we got ready to look. Considering all things, the Trustees feel a grateful surprise that this observatory with the most powerful telescope is an accomplished fact at all."[25] The Lick Observatory, he repeated, would soon speak for itself in the world of science and to the honor and fame of the University of California. Mastick, speaking for all the trustees, emphasized that the construction of such a large telescope involved many new problems and devices, and "before anything was determined upon, Floyd made himself thoroughly acquainted with that which was to be done, and the result which was to follow. He knew the point to be attained, and he sought and obtained the best means." He went on, "The Observatory, as it now is, was evolved from the knowledge derived by him from the most eminent astronomers, opticians, and astronomical mechanicians in the world, all of whom have freely and earnestly given their opinions and advice based upon their investigations and experience." In conclusion he stated, "The trust imposed upon us has been finished, and we believe that as this great instrument explores the sidereal expanse, it will develop new and wonderful things to man, thus exalting his knowledge, and making him more fit for the uses designed by his Creator."[26]

Mastick spoke briefly on the transfer. He praised the Captain who, "by his love of science, sound judgement, devotion of his time, and with great industry, has gathered together all the knowledge and experience which existed in the world concerning the construction of Observatories and the excellencies and requirements of astronomical instruments." Floyd, he noted, had written 5,000 letters (an amazing fact Floyd himself must have found unbelievable). He had received some 3,000 – and the result was "this noble Observatory." Afterward Mastick wrote to Floyd, "So you see the load is off your shoulders and ours." [27]

On July 28, 1888, a San Francisco newspaper ran an editorial entitled "California's Astronomer." [28] "There is no reason on earth why California should not have every good thing under the sun," the writer bragged. "We've got the best of climates, the richest of lands, the fairest of women, and the most public spirited of men. We have got much that money cannot buy, and there is no reason why we should not have all that it can. We have got the biggest thing in telescopes that the world knows of, and there is every reason why we should have the biggest and best kind of an astronomer to operate it. Its discoveries will amount to but little, unless they are accurately read and scientifically expounded. The Lick Observatory, as an adjunct to the State University, ought to subserve practical ends. With its aid California ought to produce a generation of young astronomers of a high order of ability. Unless that results, the Lick Observatory will promote expense rather than knowledge. The astronomer in charge ought to be a man of first-class attainments. He should possess the confidence of the scientific world. . . . Is Professor Holden such a man? We fear not."

As a result of Floyd's and Newcomb's letters the Lick trustees refused to pay Holden's bill. Holden was furious. He wrote first to the Trustees, then to T. Guy Phelps of the regents, but again his claims were turned down as groundless. He did not bring the suit he threatened. He probably realized he was fighting a losing battle, and this went against the grain of an army man.[29]

A reporter headlined the result, "HIGH PRICED STARGAZER MEETS WITH WELL DESERVED REBUFF. His bill for $6,000 for advice to the Lick Trustees Disallowed." [30] So the "Mahatma" of Mount Hamilton, as he would often be called, "in superior contact with the Mysteries of the Universe," began his reign. The reporter added, "The modesty of genius is not well developed in the intellectual, and moral make-up of Professor E. S. Holden." [31]

Meanwhile Cora Floyd thanked Newcomb for his warm-hearted concern for the Captain and said he was improving slowly. She enclosed a letter from Floyd to Mastick in which he assured Newcomb that he would be given "justice" on his role in building the Observatory "should Holden bring the suit he threatens." [32] "The Trustees will take care that the whole truth will be ventilated before the Courts." To counter Holden's claims Floyd sent Mastick 106 more letters from the Lick Trust files emphasizing Newcomb's role as principal adviser from the beginning of the project, "especially so in all questions of paramount importance, including the question of what kind of telescope should be developed." [33]

Commenting on the small difficulties with the telescope and the major ones with Holden, Floyd wrote, "A telescope, especially a great one, is *never* finished for an astronomer. It can only ever be finished by exhausting or arresting the fund for its construction. . . . The L. O. as it is, and without further fooling with, will furnish splendid original work for all the astronomers it can accomodate for a generation to come." [34]

For the first time perhaps, Holden, now in full charge of the observatory, realized the demands of the job and the wide range of skills needed under difficult conditions that persisted even after he became director. He wrote, "The ingenuity of the extraordinary beings of Jules Verne's stories would be severely taxed to meet the numerous exigencies of a year. One should be a farmer, gardener, millwright, carpenter, machinist all in one." [35]

In the *Observatory* Holden wrote of the ending of the observing season and the beginning of the "trying winter." He reminded his readers of the serious obstacles and discomforts of life on the summit of Mount Hamilton in the winter snows. "How is the food for the colony to be brought over a road which no longer exists? Where is the fuel to be had? How are city servants to be induced to share the enthusiasm of the observers?

Just at present there is no water for any domestic uses, except such as has passed through the hydraulic engines. Astronomers can stand the drawback of a film of machine oil over their drinking water, but how about the rest of the community?"[36]

He would write to the regents' secretary, J. H. C. Bonté, "This is rather a queer place for labor. It is very isolated and not much fun going – and the men get weary, not having either interests or variety. Then, too, we are all more or less like the horses of these mountains! We get 'loco,' as they say." At such an outpost even the janitor's job was unique: "He should have the manners of a lord, so as to please the visitors; he should be a good boxer, to keep order among the toughs; he should know considerable astronomy, so as to take care of the instruments, ditto chemistry – for the batteries – ditto arithmetic – to take care of the meteorological observations. He must be a good housekeeper, always in good temper, understanding how to chop kindling and to take care of our library etc. etc. – and all this for $60 per month."[37]

Now, too, the small astronomical staff began regular work. One of the first, and most eager, was Barnard. With a 6-inch portrait lens, he obtained some remarkable photographs of the Milky Way that promised to increase understanding of our galaxy. Besides myriads of stars, he observed "vast and wonderful cloud forms, with all their structure of lanes, holes, and black gaps, and sprays of stars. They present to us these forms in all their delicacy and beauty, as no eye or telescope can ever hope to see them."[38] On his plates he discerned dark clouds of matter that prevent our seeing the stars beyond. Lick astronomers later showed that this "interstellar matter" is found throughout the Milky Way system and accounts for an important part of the material of the universe.

On September 9, 1892, on the first night he was allowed to use the 36-inch telescope, Barnard became a hero overnight when the keen-eyed astronomer, discoverer of comets, found the faint fifth satellite of Jupiter.[39] Burnham, back in Chicago after resigning from the Lick Observatory, called it the greatest astronomical achievement of the century and said it would cause the world of science to ring.

As the Lick astronomers looked for wider fields to conquer, Keeler noted, "I am sure we can make it lively for the stars."[40] Indeed they did in those exciting days when the Lick telescope began to fulfill Lick's dreams, revealing new frontiers far greater than those opened by the forty-niners in their quest for gold.

The regents, some of them forty-niners who had arrived in California forty years earlier seeking treasure, were now given the chance to see neighboring worlds in our own solar system and explore distant galaxies of which they had never dreamed.

On Saturday nights the observatory was open to all who wished to look through the great telescope. Visitors arrived by "shoals." They came from every state of the Union and from many nations. In general they were intelligent, interested in all they saw, and enthusiastic over the beauty of Mount Hamilton. Amazed at the immensity of the telescope, they marveled that an instrument that weighed tons was so perfectly balanced that it could be handled with ease. Impressed by the gigantic steel dome, they wondered that it could revolve noiselessly at the turn of a hand wheel, while a similar wheel raised or lowered the circular lifting floor seventeen feet. One visitor commented, "It sometimes seems like living in a colony of magicians who deal in the mysteries of heaven and earth. . . . The processes seem exceedingly mysterious, especially in the big dome at night, dark except for the astronomer's lantern, which casts queer shadows, and the sizzling electric spark."

In the smaller 12-inch dome, Barnard often acted as a guide to those mysteries. Hour after hour he patiently answered visitors' questions, showing them the moon, planets, double stars, and other objects through the telescope. A reporter gave a vivid picture of the scene: "Here in the gloom intensified by a single 'bull's eye' hand lamp, perched up on the last round of a high observing chair, sits a young man, whose face – as an occasional lantern partly illumines the room – appears almost grand in its honest devotion to a beloved science. Here he is King, and all the interested audience, who speak in whispers and creep about as though fearful of dispelling an illusion, are most faithful subjects. Nothing tires his patience. The same questions are asked over and over again, and answered kindly and carefully. . . . When the last carriage rolls down the mountainside, he locks himself in his 12-inch dome, with a record book and chronometer and studies until dawn breaks over the snow-capped Sierras."[41]

Astronomers, too, came from every part of the world. One of the most pressing requests was from a San Franciscan, Rose O'Halloran. Thomas Bishop, a friend of Floyd's, asked him to fix a time for this "very deserving young lady" to come up the mountain. "She will want so much to be able to take a good look at the stars through the big telescope and will write up the whole thing. *No one will ever look through that great instrument who is more thoroughly devoted to the science.*"[42] Bishop wrote that she had starved herself and slept on a cot in a room without even a carpet for years to save money to get a telescope for herself. "You will find her a little queer – a pity there are not more such earnest people in the world."

George Ellery Hale, later famous for building observatories, arrived on his honeymoon in 1890. He described his visit afterward: "Darkness had fallen and no light except the little sputtering spark on Keeler's spectroscope revealed the interior. Far above our heads, outlined against the sky through the opened shutter, the long tube reached up toward the heavens. At its extremity the 36-inch object glass gathering light from the remote nebula at which the telescope was directed concentrated it in a sharp image 57 feet below. Here, after passing through the slit of the spectroscope, it was analyzed into its component lines. Their position, when measured with reference to the standard lines of the comparison spark, gave with unequalled precision, the composition and the radial velocities of the planetary nebulae. Meanwhile the driving clock carried the great tube steadily from east to west compensating for the earth's axial rotation and holding the small nebula accurately in place on the narrow slit.[43]

"The impression made by this striking demonstration of what modern engineering in the hands of a master, can do for astronomy has been intensified by time. It had been strengthened at the Lick Observatory by discussions with such leading astronomers as Keeler, Campbell, Burnham and Barnard, with whom I was to be intimately associated in later years, and by the realization that the work on nebulae we saw in progress was and still remains one of the classics of research." [44]

Hale had ordered from Warner and Swasey the mounting for his own 12-inch telescope at Kenwood. After that telescope would come the great 40-inch refractor at Yerkes Observatory in Wisconsin, the 60- and 100-inch reflectors on Mount Wilson, near Pasadena, and the 200-inch reflector on Palomar Mountain, above San Diego, all in a way inspired by the pioneer 36-inch telescope on Mount Hamilton. How delighted Floyd would have been to be on Mount Hamilton to welcome the twenty-two-year-old Hale and share his vision of the future!

In June 1887 Floyd had heard rumors of a large observatory to be built on Wilson's Peak, above Pasadena in southern California. Edward F. Spence was said to have given $50,000 for a 40-inch telescope, to "lick" the Lick 36-inch. That summer Floyd received a letter from President M. M. Bovard of the University of Southern California, asking him for a "plan" for a large observatory. Fraser and Floyd worked out a proposal; Floyd approved it and sent it to Bovard, who thanked him for his kindly interest in the enterprise.

On August 21, 1887, Floyd was surprised to receive a heart-warming letter from his old friend Barty Shorb, the son-in-law of Don Benito Wilson, for whom Wilson's Peak was named. Shorb wrote from the San Gabriel Wine Company in southern California. "My dear Dick, I find a moment to devote to my social life and my thoughts revert to you. I don't know how to apologize to you my dear friend for all my seeming neglect of you. My life for many months past has been such a miserable turmoil of business cases that half of the time I am driven almost to distraction. You know how dearly I love you and how much I appreciate your society how pleasant it is to turn from the business friends plotting to rob you, to the gentleman and friend as you have always been to me. Remember this, dear Dick, no man in the State has so entirely my heart and respect than yourself and it would distress me beyond measure to feel that our friendship which has existed over so many years should grow less or strained."[45]

Then came the surprising part as Shorb wrote of the observatory on Wilson's Peak. He said he was waiting to see a "party" who was away. "The moment he returns I will immediately lay the whole matter before him, and I can then write something definite. We can raise here $100,000 for the Observatory, and I would like to have it under your direction and consideration. Could you do that for us?" We do not know if Floyd ever considered his friend's suggestion seriously. Certainly, at this time, he longed only for the peace of Kono Tayee.

Months passed, and Floyd heard nothing more until, in October 1888, Fraser wrote to say that after a year with Miller and Lux, he was going to Banning to work on his own place there. "The chances is that I will construct that Observatory for the 40-inch glass from Bovard and Spence of Los Angeles. What do you think of it?" He had had some correspondence with Bovard and wanted to talk it over with Floyd before going south. "The mountain where it is going is about as high as your place but I think it is freer of fog."[46]

Early in 1889 Fraser sent on a copy of the *Pasadena Union,* with accounts of the expedition to Wilson's Peak. In addition to Captain Thomas Fraser, it said, the party included astronomer W. H. Pickering, Judge B. S. Eaton, an old mountaineer, and President Bovard. On the peak the climbers found an old cabin filled with snow. They shoveled it out, and that night Judge Eaton regaled the group with hair-raising tales of bears prowling those mountains. The following morning Fraser, asked his opinion of the site, said he thought the location excellent – in most respects superior to Mount Hamilton. He then surprised the party by launching into a speech and unfurling a flag, bordered by a fringe of colored silk, and presenting it to President Bovard. He talked of James Lick and the beginnings of his observatory and of the choice of Mount Hamilton as the site. "When we began to talk about it the people called us humbugs, but the people of California are not humbugs. When they take one step in a worthy enterprise they will take another. When I first went to

Mount Hamilton I felt sure something would be done, and I feel equally sure something will be done here."[47]

That night a banquet sponsored by the Board of Trade was held at the Carlton Hotel in Pasadena. Fraser, in his element, had composed a poem dedicated to E. F. Spence. He spoke eloquently of the "absolute" need of a telescope on Wilson's Peak and urged the building of a road.

But Spence's dream vanished. The plan for an observatory on Wilson's Peak exploded with the bursting of the Los Angeles land bubble. The 40-inch lenses would go instead to the observatory at Williams Bay, Wisconsin, founded by another tycoon, Charles T. Yerkes, and built by George Ellery Hale.[48] But even before the fate of the lenses was decided, Tom Fraser had left for another "sphere." If he had lived, he might have played a leading role with Hale in the building of the Mount Wilson Observatory on this same peak fifteen years later.

Another guest at the Carlton banquet was the optical expert John Brashear. After visiting Keeler on Mount Hamilton, he wrote to Floyd from the Palace Hotel in San Francisco, "I wish I could tell you how much I admired the grand work you have done at the Lick Observatory; but sub rosa I cant say so much of your director for all he treated me with marked consideration, but my dear Captain your humble serv't though a little dull of comprehension had no difficulty in seeing that it is only 'skin deep.' He is the great I AM [written in huge letters]. Yet I know that Keeler is his peer if not his Superior any day." Brashear longed to make the trip to Kono Tayee – but it was impossible. "Oh," he exclaimed, "how I would like to have a chat with you in your good old fashioned way, hear your good laughs and your splendid jokes."[49]

On New Year's Day, 1889, Floyd got up early to be ready to observe the total eclipse of the sun; it was a magnificent morning. For several days before, it had stormed and rained "like the devil." It had been hard to make the necessary preparations. The 5-inch objective glass he had ordered from Alvan Clark had arrived in December, along with the tube, declination axis, and "eye end arrangement." He had to make the balance of the wood mounting himself. He was delighted with the instrument, which by "capsizing" the crown lens, he could change from a visual to a photographic telescope.[50]

From his weather log, kept since 1873, he estimated chances for a clear day to be about even. But he had bet astronomer David Todd a case of champagne it would be clear. "California gave us a magnificent day for the occasion." With his roughly arranged little telescope, without a driving clock, he managed to get a couple of random shots, "unskilfully made" because, he said, he was "much out of practice in this kind of thing," and his health did not allow him to "fool around" much in the dark-room. Still, as nothing depended on his work, even as a sick man he managed to get about four weeks of pleasant occupation out of the eclipse. He sent a couple of his shots to Newcomb, thinking they might interest or amuse him, and added, "I wish you could come out here and have a rest with me. I would cruise you around on the Lake and roll you around on a Tricycle."[51]

The Lick Observatory eclipse party had been stationed at Bartlett Springs, nine miles north of Kono Tayee. On his way back, Keeler stopped off for a couple of days to visit Floyd. After the "villainous fare" on the eclipse expedition, he welcomed the fine meals and talk of the good old days on the mountain. The trip from Sites, the station nearest the Springs, was made in an open wagon through deep mud to the croquet court of the hotel, where Barnard dug a pit for the telescope. They too had had days of rain but the morning of January 1 dawned clear. Keeler's spectroscopic observations showed that the features of the corona were independent of the moon's position against the sun and were not caused by diffraction. Barnard obtained the most "exquisite" photographs of the eclipse, showing many structures in the corona; the image was only half an inch in diameter.[52] Keeler praised Floyd's photograph of the corona, and Burnham, when he saw it, said it was not beaten by any he had seen.

Months passed, and Fraser continued to worry about the Captain. A year later he wrote, "I think it is about time I was wrighting you a letter. I have so many things to say that it seems impossible to say them in a letter but would like to have one of our old fashioned chats with you." He was plodding along, happy to be working for himself. "We Run this little Hotel My Wife does most of it and I am fixing the place up and doing some farming. . . . I have put in alpha [alfalfa] wheat Potatoes and corn I have also put out some figs olives and shade trees so you see I have been busy." [53]

To this he added, "I made a trip with Clark and Pickering to Wilson's Peak and they have been wanting me to take hold with them to construct the Observatory. They will not build it for 3 or 4 years and I dont know as I want any more observatorys in mine. I am like you I want to be let alone and have nothing more to do with crazey astronomers." [54]

He was glad to see that Holden had been "sat down on. . . . All the Regents but Rodgers is down on him and Rodgers is like himself only a small part of a man. I doubt now if you saw the San Jose newspaper where Dr. Shorb spoke so well of you thank god you have friends that can Remember what you done for the Observatory – Enough of that Observatory business I know it bores you." [55] Trying to cheer his old friend, he concluded, "If there is a war with Germany I want to go to sea with you in a Privateer. Please write soon and tell me what you are doing."

In those days Floyd looked forward to working in his little observatory. Soon, however, he became too ill to see anyone. In one of his last letters to his "Dear Hal," he told her of the *Whisper,* a little steamer built for him by his friends in San Francisco and Lakeport. "I took your mama with me in a carriage today and went to the Union Iron Works to see the *Whisper.* I was much pleased with her and think she will be a fine little boat." [56] When the steamer was delivered at Clear Lake, its Captain was no longer there.

After consulting doctors in San Francisco, Cora, desperate over the Captain's condition, decided to go with him to Philadelphia to consult a heart specialist. But it was too late. On October 17, 1890, Richard Samuel Floyd died at the Aldine Hotel in that city.[57] He was just forty-seven. He had given the best of those forty-seven years to the Lick Observatory. Cora, after watching anxiously over him in these last terrible months, was inconsolable. With the heart of her life gone, she felt she had little left to live for.

With seventeen-year-old Harry and James Keeler, Cora Floyd brought the Captain's body back to San Francisco in a large casket in a private Pullman car. Funeral services were held in Grace Episcopal Church, at Stockton and California streets. The twenty-four pallbearers included the Lick trustees, the astronomers Barnard, Burnham, and Keeler, and Tom Fraser, who had come from Banning to say a last good-bye to his old friend. They were joined by the university regents, representatives of the Society of California Pioneers and the California Academy of Sciences, led by Charles Crocker, and Floyd's old friend Tom Madden, with officials of the Chamber of Commerce and the San Francisco and Pacific yacht clubs. After the service a long procession of carriages followed the hearse to Laurel Hill Cemetery, at the end of Bush Street. On Floyd's grave the spectacular floral representation of the Lick Observatory on Mount Hamilton dominated the other floral tributes.

A few days earlier, at a meeting of the California Academy of Sciences, President Harvey W. Harkness had announced Floyd's death and said, "It would be only fitting if a grave on a mountain he loved so well were given him."[58] Burnham, speaking warmly of the proposal, said, "There is probably not a living astronomer who could have so successfully carried out this work and made Lick Observatory what it is today." He suggested that Floyd's ashes be placed near the body of James Lick, "with a column of the great telescope as their perpetual monument."[59] Barnard echoed Burnham's feelings on Floyd's service to astronomy in carrying to a material success Lick's noble gift, with the great observatory "a monument to the man who unselfishly gave the best years of his life in making that bequest a reality."[60] Keeler, in the same vein, spoke of Floyd's immense contribution, of his "sacrifice of personal interests which those who were acquainted with the Captain know, was always cheerfully made."[61]

The proposal to bury Floyd's ashes on the mountain was never realized. They remained at Laurel Hill until moved to the Henry Augustus Lyons Mausoleum at Cypress Lawn in San Bruno, where they are today.

If Floyd had been so honored, it would have been fitting to recognize Tom Fraser in a similar way. Just a year after his friend's death, Fraser, like Floyd, worn out by his years devoted to building the observatory, died at his home in Banning. An illness resulting from a liver infection developed into typhoid fever. He was only forty-nine. A friend called the construction of the observatory the crowning glory of his life. "To his genius and perseverance much of the success of the completion of that wonderful edifice is due. He built it, and for ages to come it will stand as a monument of Captain Fraser's achievement in his chosen vocation."[62]

This was a sad time indeed for the Floyd family and their friends, a time made darker by Cora Floyd's death on February 27, 1891, at the Hotel San Rafael, just four months after the death of the man she adored. Her death resulted from pneumonia, following a severe attack of "nervous fever" brought on by the strain of the Captain's long illness and death.[63] In a way she, too, was a victim of the stress of building the observatory.

Epilogue

So ends the saga of Captain Richard Floyd and the Building of the Lick Observatory. As Mary Lea Shane, wife of the former observatory director Donald Shane, wrote, "In all ways the Lick Observatory was a remarkable venture. From every aspect it was a pioneering feat, without pattern or precedent to follow. Lick's gift prescribed 'the most powerful telescope ever built,' in itself a unique requirement. The Observatory was the first permanent astronomical installation at a high altitude. A small residential community must be established on an isolated mountain, 4200 feet high, initially without roads or water. This work must be accomplished literally by 'manpower,' for there were few mechanical devices available and electric power was still many years away. Yet in eight years from the beginning of active work, the finest observatory of its time was built, equipped and in operation."

In the next hundred years that observatory's telescope would continue to penetrate farther and farther into space to reveal a far greater universe than James Lick or Richard Floyd could ever have imagined.

Notes

This book is based mainly on primary sources – letters, reminiscences, personal recollections, and contemporary magazine and newspaper articles. Many are in the Mary Lea Shane Archives of the Dean McHenry Library at the University of California at Santa Cruz, where they were moved from Mount Hamilton in 1966. These papers are supplemented by the Richard S. Floyd papers in the Bancroft Library of the University of California in Berkeley. Most of these came from Floyd estate, Kono Tayee, at Clear Lake. They relate, in large part, to Floyd's last years at the Lick Observatory and were given to the Bancroft by Mrs. Gustave Vouté in 1958. It is hoped that some day they may be deposited in the Mary Lea Shane Archives in Santa Cruz so that all the Floyd records relating to the founding of the Lick Observatory can be coordinated in one place.

Another primary source for this saga is the correspondence with Simon Newcomb, chief adviser to the Lick Trust and Floyd's chief correspondent in this period. Much of this correspondence is in the Manuscript Collection of the Library of Congress. Many of the letters are duplicated in the Lick Archives. These primary sources are illuminated by biographies and autobiographies such as Simon Newcomb's *Reminiscences of an Astronomer.* At the Library of Congress also are the papers of W. G. McAdoo, who married Floyd's sister Mary. These papers give insight into the background of Floyd's life and early activity in the South.

At the Bancroft Library the recollections of the Lick trustees are recorded in the H. H. Bancroft Collection. In that collection, too, are the valuable *Recollections* of James Lick's friend David Jackson Staples. Here, too, are John Knoche's *Recollections of the Early Days of a Pioneer, 1849* and the map of California from Clear Lake to Mount Hamilton, published in 1871 in the *Handbook of Calistoga Springs.*

Other important archival sources are the Henry Draper papers at the New York Public Library, the Joseph Henry correspondence in the Smithsonian Institution Archives, in Washington, D.C., and the Norman Lockyer Archives in Sidmouth, Devonshire, England.

The papers of E. S. Holden in the Mary Lea Shane Archives in Santa Cruz are voluminous for the founding period and for the subsequent period when Holden was president of the University of California and then the first director of the Lick Observatory.

Henry Mathews, Secretary of the Lick Trust, wrote many accounts of the Lick Trust. These include his *Reminiscences* (handwritten ms., 1920, SLA), supplemented by material in the Bancroft Library written when the Lick Trust was challenged by Holden (typewritten, 1888). *Some Recollections of the James Lick Trust,* written in San Francisco in March 1918, is at the Society of California Pioneers. There too is his autobiography, *Some Recollections of Eighty-seven Years of a Lifetime,* written in 1927. There also are two volumes of Mathews's scrapbook, with clippings and an article by Fraser on Floyd (pp. 122–3) in which he calls Floyd the best seaman and navigator in command of a steamer in the seventies. See also Mathews's *In Memoriam,* on Floyd. In August 1918, Mathews gave a copy of his brief *Some Recollections of the Lick Trust* to the San Francisco Public Library.

The following is a list of abbreviations used in the notes:

Archival sources

BUC — H. H. Bancroft Collection, George Davidson Papers, and Phoebe A. Hearst Papers, Bancroft Library, University of California, Berkeley
CHS — California Historical Society, San Francisco
FBUC — Richard S. Floyd Papers, Bancroft Library
HCO — Records of the Harvard College Observatory, in Harvard University Archives, in Cambridge, Massachusetts
HDNYPL — Henry Draper Papers, New York Public Library
HEH — Manuscripts Department, Henry E. Huntington Library, San Marino, California
LC — Manuscript Division, Library of Congress, Washington, D.C.
McLC — W. G. McAdoo Papers, LC. Papers of the Floyd family
NLC — Simon Newcomb Papers, LC
PFGHS — Picot Floyd Papers, Georgia Historical Society, Savannah
SCP — Society of California Pioneers
SIA — Smithsonian Institution Archives, Washington, D.C., Records of the Office of the Secretary
SLA — Mary Lea Shane Archives of the Lick Observatory, Dean McHenry Library, University of California, Santa Cruz
UCA — University of California Archives, Bancroft Library, Berkeley
USNO — United States Naval Observatory
VU — Edward Emerson Barnard Papers, Special Collections, Vanderbilt University, Nashville, Tennessee

Journals

ASP — Astronomical Society of the Pacific
BAA — British Astronomical Association
CIW — Carnegie Institution of Washington, Annual Reports
LOPUB — *Lick Observatory Publications*
MNRAS — *Monthly Notices of the Royal Astronomical Society*
PASP — Publications of the Astronomical Society of the Pacific
SM — *Sidereal Messenger*

Individuals

EEB — E. E. Barnard
JAB — John A. Brashear
SWB — S. W. Burnham
WWC — W. W. Campbell
GD — George Davidson
HD — Henry Draper
CLF — Cora L. Floyd
HALF — Harry A. L. Floyd
RSF — Richard Samuel Floyd
TEF — Thomas E. Fraser
GEH — George Ellery Hale
ESH — Edward S. Holden
JEK — James E. Keeler
WFK — William F. Keeler
MFMcA — Mary Floyd McAdoo
EBM — E. B. Mastick
HEM — Henry E. Mathews
SN — Simon Newcomb
DPT — David P. Todd
W & S — W. R. Warner and A. Swasey

Chapter 1: The story of James Lick

1 Many accounts of James Lick have been written. One of the most important is that of Willard B. Farwell. Farwell, a friend of Lick's, was asked by the Society of California Pioneers to write Lick's biography, with E. J. Molera, of the California Academy of Sciences. Many years later *The Life of James Lick,* written by Farwell, was published in the *Quarterly of the SCP,* vol. 1, no. 2 (1924), but Farwell was never given credit.

2 Much of the material in this chapter is based on Farwell's account. Unless otherwise specified, this was my primary source. Another source on "the adventurous Lick" is Oscar T. Shuck, *Sketches of Leading and Representative Men of San Francisco,* vol. 2 (San Francisco, 1875), 841–58. Also William H. Worrilow, *James Lick, Pioneer and Adventurer: His Role in California History* (1949). This is said to have been written by Arthur D. Graeff.

3 According to Edouard Stackpole of Nantucket, the *Lady Adams,* the brig on which Lick sailed, was built in Baltimore in 1825.

4 Interesting accounts of San Francisco and its waterfront at the time of Lick's arrival in San Francisco are B. E. Lloyd, *Lights and Shades in San Francisco* (San Francisco, 1876), 91–7; for a description of San Francisco in this period, see Samuel Williams, *The City of the Golden Gate: A Description of San Francisco in 1875* (San Francisco, 1921). Also Kevin Starr, *Americans and the California Dream: 1850–1915* (New York, 1973).

5 An account of his journey to the gold fields is found in W. F. Swasey, *The Early Days and Men of California* (Oakland, Calif., 1891), quoted by Howard Sonenfeld in his valuable M.A. thesis for the University of Pennsylvania in 1937. A copy was donated to the Bancroft Library on July 22, 1958. Its title is *James Lick and the History of California.*

6 Farwell, *Lick.*

7 "The Story of the Lick Mill" was told by Thomas E. Fraser to Simon Newcomb in 1885. Today it is in the Mary Lea Shane Archives at the University of California at Santa Cruz.

8 Another account of Lick's life is given by Eugene T. Sawyer, *History of Santa Clara County, California* (Los Angeles, 1922).

9 Farwell, *Lick,* 39–40.

10 John Knoche, ms. 141 in BUC.

11 Lick to Conrad Meyer, Dec. 17, 1872, quoted by Farwell, *Lick,* 39–40.

12 Quoted by Farwell, *Lick,* 38.

13 John S. Hittell, *History of California,* vol. 4 (San Francisco, 1897), 577–82.

14 The great Hall of Mirrors provided Lick with the inspiration for the Lick House dining room. Sonenfeld, *James Lick,* 5, 54.

15 Amelia Neville Scrapbook, CHS, 190–1.

16 *San Francisco Daily Examiner,* June 1893.

17 Sonenfeld, *James Lick,* 62.
18 Ibid.
19 G. W. James, *How We Climb to the Stars and the Lick Observatory* (San Francisco, 1887), 17–18.
20 Rosemary Lick, *The Generous Miser: The Story of James Lick of California* (Menlo Park, Calif., 1967), 62.
21 Farwell, *Lick,* 42.
22 ESH, *Handbook of the Lick Observatory of the University of California* (San Francisco, 1888), 12.
23 Madeira to ESH, July 14, 1887, SLA.
24 Joseph Henry, on his meeting with Lick, SIA. (Henry was in San Francisco Aug. 23 through Sept. 18, 1871.)
25 Louis Agassiz, *Proceedings of the California Academy of Sciences* 4 (1872), 253–6.
26 Staples, account of meeting with Lick, H. H. Bancroft Collection, BUC.
27 Ibid.
28 Farwell, *Lick,* 15, 30.
29 Sonenfeld, *James Lick.*
30 Staples, *Lick meeting,* BUC.
31 The story of the University of California is told in W. W. Ferrier, *Origin and Development of the University of California* (Berkeley, 1930); V. A. Stadtman, *The University of California 1868–1968* (New York, 1970); and W. C. Jones, *Illustrated History of the University of California* (Berkeley, 1901).
32 Staples, *Lick meeting,* BUC.
33 ESH, *LOPUB,* 1 (1887).
34 GD in *U.S. Coast and Geodetic Survey Report* (1872), 172–6.
35 GD to Benjamin Peirce, Feb. 1, 1869, letter marked Private, National Archives (R.G. 23) Supt.'s file; also HCOA, Widener Library.
36 *Proceedings of the California Academy of Sciences* 5 (1873–4).
37 TEF "Biography," *Banning Herald,* Oct. 10, 1891, SLA.
38 Interview with Lovett Fraser at Lakeport in 1959.
39 John Fraser biography in Aurelius O. Carpenter and Percy H. Millberry, *History of Mendocino and Lake Counties* (Los Angeles, 1914), 317–20.
40 TEF, "Biography," *Banning Herald.*
41 Lick to Henry, Oct. 22, 1873, SIA and SLA.
42 Henry to Lick, Aug. 13, 1874, SIA.
43 Henry to T. H. Huxley, Aug. 1, 1874, SIA.
44 Henry to Lick, Aug. 16, 1874. A good account of this period is given in Howard S. Miller, *Dollars for Research* (Seattle, 1970), chap. 5.
45 Henry to Lick, Aug. 13, 1874, SIA.
46 GD to SN, Nov. 11, 1873, NLC.
47 *Scribner's Monthly* 7 (1873–4), 46.

48 SN, *Reminiscences of an Astronomer* (Boston, 1903), 183.

49 SN, *Harper's Magazine*, Feb. 1885.

Chapter 2: Captain Floyd and the Lick Trust

1 TEF, *Report*, 1876, SLA.

2 Ibid.

3 On the Floyd background, see James Vocelle, *History of Camden County, Georgia* (n.p., 1914; reprint, Kingsland, Ga., 1967); and *Camden's Challenge: A History of Camden County, Georgia* (Camden County Historical Commission, 1976), E. Bailey, chairman; Marguerite Reddick, compiler; Eloise Bailey and Virginia Proctor, editors.

4 The best account of the *Florida* is Frank Lawrence Owsley, Jr., *The C.S.S. Florida: Her Building and Operations* (Philadelphia, 1965); National Archives, Navy Records, ORN Group 45.

5 Ibid.

6 Floyd recalled some of his adventures on the *Florida*, in a letter to Captain John W. Morton of Nashville, Clear Lake Copy Book, 1885, 215; RSF to Col. J. D. Porter in Paris, Tenn., Jan. 21, 1870, FBUC.

7 Owsley, *Florida*.

8 RSF to his sister, Mary F. McAdoo, June 15, 1865, from Paris, France, McLC.

9 John Fraser biography in Aurelius O. Carpenter and Percy H. Millberry, *History of Mendocino and Lake Counties, California* (Los Angeles, 1914), 317–20; RSF to MFMcA, Sept. 7, 1866, McLC.

10 Biography of Henry Augustus Lyons in J. Henry Johnson, *History of the Supreme Court Justices of California*, vol. 1 (San Francisco, 1961).

11 Cora Lyons's background in Louisiana can be traced in Edwin Adams Davis, *Plantation Life in the Florida Parishes of Louisiana, 1836–1846* (New York, 1943); Herman de Bachellé Seebold, *Old Louisiana Plantation Houses and Family Trees* (New Orleans, 1941).

12 Clipping on Floyd wedding in CLF scrapbook, CHS.

13 Amelia Neville, *The Fantastic City: Memories of the Social and Romantic Life of Old San Francisco* (Boston, 1932), 75, 138–9.

14 For a good account of shipping in this period, see H. Parker and Frank C. Bowen, *Mail and Passenger Steamships of the Nineteenth Century* (Philadelphia, 1927); Gordon Newell and Joe Williamson, *Pacific Coastal Liners* (Seattle, Wash., 1959).

15 CLF to her sister at Oakley Plantation, St. Francisville, La., Feb. 6, 1873, Oakley archives, St. Francisville, and SLA.

16 James Lick to Lick trustees, March 24, 1875, SLA.

17 Fraser, *Report*, 1876, SLA.

18 James Lick to William Lick, March 29, 1859, HEM Scrapbook, SCP.
19 Most of this account is based on Farwell, *Lick.*
20 Extract from the *Napa Reporter* in CLF Scrapbook, CHS.
21 H. C. McDaniel to RSF, July 29, 1877, FBUC.
22 Mathews, *Reminiscences,* SLA.
23 Mathews, *Some Recollections,* SCP.
24 Mathews became secretary of the Trust in January 1876.
25 Farwell, *Lick.*
26 TEF, *Report,* 1876, SLA.
27 William H. Brewer, *Up and Down California in 1860–1864,* ed. Francis P. Farquhar (New Haven, Conn., 1930); W. H. Brewer to A. T. Denny in San Francisco, from New Haven, Jan. 3, 1888, SLA.
28 TEF, *Report,* 1876, SLA.
29 *Harper's Weekly,* Oct. 23, 1875. The height of Mount Hamilton is now given as 4,200 feet.
30 ESH, *LOPUB,* vol. 1, 1888.
31 The University of California (formerly the College of California, founded by Henry Durant) moved to its present site in Berkeley in 1873.
32 RSF to SN, March 4, 1878, SLA.
33 RSF to the University of California regents, Nov. 1, 1875, SLA.
34 University of California regents to Lick trustees, Dec. 7, 1875, SLA.
35 Benjamin Alvord to ESH, Nov. 20, 1875, NLC.
36 Oscar Lewis, *George Davidson: Pioneer West Coast Scientist* (Berkeley, 1954).
37 See L. L. Paulson's *Handbook and Directory of Napa, Lake, Sonoma and Mendocino Counties* (San Francisco, 1874), 39.
38 RSF to Marmaduke Hamilton, July 21, 1879, SLA.
39 James Lick to Santa Clara Board of Supervisors, Sept. 4, 1875, SLA.
40 "History," *San Jose Daily Mercury,* June 28, 1888, SLA.
41 A. T. Hermann to RSF, mid-March, 1876, SLA.
42 RSF to Senator Aaron Sargent, April 1, 1876, SLA.
43 *Argus,* Jan. 11, 1877.
44 TEF to RSF; TEF to HEM, April 29, 1876, SLA.

Chapter 3: European journey

1 S. P. Langley, *The New Astronomy* (Cambridge, Mass., 1888), 123.
2 Simon Newcomb, *Reminiscences of an Astronomer* (Boston, Mass., 1903), chap. 1; for Newcomb, see also Arthur L. Norberg, "Simon Newcomb and 19th Century Positional Astronomy," Ph.D. diss., University of Wisconsin, 1974.
3 SN at dedication of Yerkes Observatory, Oct. 22, 1897.

4 SN to D. O. Mills, Oct. 8, 1874, SLA.
5 As Newcomb pointed out to Mills, "The difficulty of attaining perfection increases with the size of the glass." Ibid.
6 Newcomb had left for Europe at the end of December 1874. He submitted his report to the Lick trustees on March 4, 1875.
7 RSF to John Nightingale, May 23, 1876, SLA.
8 D. O. Mills to ESH, Oct. 19, 1874, SLA.
9 J. Henry to Lick, Dec. 13, 1873, SIA.
10 Herbert Dingle, quoted by R. G. Aitken in *Driving Back the Dark,* PASP Leaflet No. 101, June–July, 1937.
11 SN to F. Ward, Secretary of the Lick Trust, Oct. 14, 1875, SLA. See also HEM, *History,* FBUC.
12 Henry Rowland to J. Clerk Maxwell, in Nathan Reingold, ed., *Science in Nineteenth Century America* (New York, 1964), 269.
13 HEM on RSF at Grubb's, March 1876; RSF on Grubb. For a biography of the Grubbs, see The Reyrolle Grubb Parsons Co. Ltd., *Two Fathers and Two Sons* (Newcastle-upon-Tyne, 1971).
14 Reingold, *Science,* quotes J. W. Draper to Henry Draper, Dec. 1, 1870.
15 SN *Report,* March 4, 1875; *Operations under the Provisions of Mr. Lick's First Deed of Trust, LOPUB,* vol. 1.
16 Grubb to RSF, June 14, 1876, SLA.
17 RSF on Copeland, quoted by HEM, *Reminiscences,* SLA.
18 Grubb to RSF, July 29, 1876, SLA.
19 C. E. Mills and C. F. Brooke, *A Sketch Life of Sir William Huggins* (London, 1936), 25.
20 Huggins on Floyd's visit, July 26, 1876. Also *CIW Annual Report* 1 (1902).
21 William Lassell to Henry Draper, March 13, 1877, HDNYPL.
22 SN to D. O. Mills, Oct. 8, 1874; SN, *Report.*
23 SN to ESH, July 2, 1876, NLC. Actually Holden went with E. B. Knobel to Scotland on a yachting tour. *MNRAS* 75(1914):264; ESH to SN, July 20, 1876, FBUC.
24 RSF to EBM, June 27, 1888, SLA. But the story that he had traveled with RSF to various observatories was perpetuated by ESH, Milicent Shinn, and others, e.g., ESH in *Overland Monthly,* Nov. 1892.
25 *Nature,* July 6, 1876.
26 SN to ESH, July 25, 1876, NLC.
27 *Nature,* July 13, 1876.
28 SN to ESH, July 25, 1876.
29 SN to D. O. Mills, Oct. 8, 1874, SLA.
30 SN, *Reminiscences,* 328.
31 Sir William Thomson at the BAA meeting in Glasgow that began on Sept. 6, 1876.
32 Ibid.

33 Grubb to RSF, Aug. 10, 1876, SLA.
34 Gill to ESH, Aug. 19, 1876, SLA.
35 Gill to RSF, Sept. 18, 1876, SLA. The Irish astronomer, Thomas Robinson, shared Gill's view that only with large reflectors could "the utmost depths of celestial exploration in stellar spectroscopy . . . and nebulae be reached." "On the Relative Power of Achromatic and Reflecting Telescopes," *MNRAS* 36(1876):307. This article appeared while Floyd was still in Europe.
36 Gill to RSF, Sept. 18, 1876, SLA.
37 RSF to Faxon D. Atherton, Aug. 24, 1876, SLA.
38 Charles Plum, interview with Lick recorded in H. H. Bancroft Collection, 1886, BUC.
39 Ibid.
40 *San Jose Daily Mercury,* 1888, typed copy, SLA.
41 Some said John was wise to refuse to sign, as he was afraid that thereby he would be signing away any future rights he might have to the property.
42 HEM, *Reminiscences,* SLA.
43 Ibid.
44 Willard Farwell, *Lick,* 58.
45 HEM, *History,* FBUC; Lick told Staples the same thing.
46 Interview with Plum, H. H. Bancroft Collection, BUC.
47 Plum, interview, BUC.
48 *Napa County Reporter,* Jan. 3, 1874.
49 Lick was at Calistoga in June 1875.
50 Biographical information on Mastick, H. H. Bancroft Collection, BUC. When Mastick died, his astronomical library was given to the Astronomical Society of the Pacific.
51 Biographical information on Sherman, H. H. Bancroft Collection, BUC.
52 Leverrier died Sept. 23, 1877, SLA.
53 RSF to C. Feil, Oct. 4, 1877, SLA.
54 Agnes Clerke, *A Popular History of Astronomy During the 19th Century* (London, 1902), 171.
55 RSF to HEM, Sept. 25, 1876, SLA.
56 Grubb to RSF, Sept. 25, 1876, SLA. Grubb described this telescope in *Engineering* 29 (1880).
57 RSF to Grubb, Dec. 19, 1879, SLA.

Chapter 4: Transition to the skies over Mount Hamilton

1 Schönewald to RSF, Oct. 1, 1876, SLA.
2 RSF to HEM, Oct. 6, 1876, from Nice, SLA.

3 Ibid.
4 HEM to Charles Zeile, quoted by Rosemary Lick, *Miser,* 76.
5 Newspaper accounts, SLA.
6 RSF, on Lick's desire to be buried under the great telescope, Sept. 3, 1890, FBUC.
7 Oscar Lewis, *Davidson.*
8 RSF, on Lick's tomb.
9 EBM on RSF's travels for the Lick Trust, Dedication, June 27, 1888. Also SN, *Report.*
10 Bills in FBUC.
11 RSF to HEM, Jan. 23, 1877, SLA.
12 RSF to HEM, April 10, 18, 1877, SLA.
13 Guillermo Prieto, *San Francisco in the Seventies: The City as Viewed by a Mexican Political Exile* (San Francisco, 1875), 76.
14 On Nov. 29, 1876, the Second Board was retired, and the Third Board was formally elected.
15 Court Agreement, Jan. 14, 1878, SLA.
16 Statement to the Finance Committee. The trustees applied to the state legislature of 1878 and offered to pay $15,000, the face of the assessment. The legislature accepted the offer. Sonenfeld, *James Lick,* gives a good account of the transactions on the Lick estate, BUC.
17 Advertisement, Nov. 17, 1874, SLA.
18 Cemetery records at Cypress Lawn Cemetery, Colma.
19 *San Jose Mercury,* May 19, 1878, SLA.
20 Taliesin Evans, "A Californian's Gift to Science," *Century Magazine* 32 (May 1886):70.
21 *San Jose Mercury,* May 19, 1878, SLA.
22 See Olin Eggen, *S. W. Burnham,* PASP Leaflet no. 259; EEB, "Sherburne Wesley Burnham," *PA* 39(1921):309–24.
23 TEF, "log," June 25, 1879, SLA.
24 *Report to the Trustees of the James Lick Trust of the Observations Made on Mt. Hamilton with Reference to the Location of the Lick Observatory* (Chicago, 1880); Report by SWB, *LOPUB* 1 (1888).
25 Ibid.
26 SWB to RSF, Sept. 9, 1879, SLA.
27 Newcomb took charge of the Nautical Almanac office on Sept. 18, 1877.
28 RSF to TEF, Sept. 10, 1879, SLA.
29 Ibid.
30 SN, *Reminiscences,* 188.
31 SN to RSF, Oct. 18, 1879, SLA.
32 Ibid.

Chapter 5: "Dear Captain"

1 RSF to SN, Jan. 30, 1880, NLC.
2 RSF to SN, May 9, 1879, SLA.
3 Ibid.
4 Charles Plum, on visit to Grubb, July 29, 1879, SLA.
5 Joseph Henry had proposed Watson, saying he knew of no one in the country he could recommend more highly; July 13, 1877, SLA.
6 TEF to HEM, May 12, 1880, SLA.
7 ESH to T. G. Phelps, Sept. 9, 1892, SLA.
8 WWC, *Holden,* 8, 1919.
9 ESH to SN, Nov. 22, 1875, SLA.
10 SWB to ESH, Feb. 20, 1889, SLA.
11 ESH to HD, Jan. 8, 1875, HDNYPL.
12 ESH to HD, Sept. 4, 1877, HDNYPL.
13 SN, *Reminiscences,* 184.
14 ESH to D. O. Mills, Oct. 8, 1874, SLA.
15 A. W. Bowman to ESH, Nov. 11, 1874, SLA.
16 ESH to SN, Nov. 22, 1875, NLC.
17 ESH to HD, Oct. 15, 1875, HDNYPL.
18 William Alvord to ESH, May 17, 1875, SLA.
19 Henry to Lockyer, Aug. 4, 1874, Norman Lockyer Observatory, Sidmouth, England; also in SIA.
20 SN to candidate (name illegible), Oct. 20, 1875, NLC. Even as late as 1882 the problem had not been solved; SN wrote then to J. P. Esty at Amherst to ask how David P. Todd would get along as director of a large observatory; March 15, 1882, NLC.
21 SN, *Reminiscences.* ESH to SN, Oct. 15, 1875, HDNYPL. They discussed the organization and building of the observatory. Afterward Newcomb asked Holden to write an account of the discussion. Long after, as we shall see, Holden would claim entire credit for the original observatory plans. SN to D. O. Mills, Oct. 8, 1874, SLA.
22 TEF, *Report,* 1876, SLA.
23 TEF and RSF in Washington, TEF, "Log," 1880, SLA.
24 Alvan Clark, "Autobiography," *SM* 8 (1889):109–17.
25 Clark to Lick, Nov. 3, Dec. 16, 1873, SLA.
26 Clark to Lick, June 25, 1874, SLA.
27 Grubb to RSF, Feb. 12, 1881, SLA. From the beginning Clark had set his heart on the making of the big glass. See also Deborah Jean Warner, *Alvan Clark & Sons: Artists in Optics* (Washington, D.C., 1968).
28 HEM to SWB, June 21, 1880. The plans were worked out by S. E. Todd.

29 TEF, "log," July 20, 1880, SLA.
30 Her letters to Tom written when she was in the seminary are in the SLA.
31 TEF, "Biography," *Banning Herald,* Oct. 20, 1891, SLA.
32 Fraser to William Sherman, Aug. 2, 1883, SLA.
33 TEF, "log," Aug. 1880, SLA.
34 TEF, Report to the Lick Trustees, Sept. 22, 1880, SLA.
35 TEF, "log," Sept. 17, 1880, SLA.
36 TEF, "log," Oct. 22, 1880, SLA.
37 TEF to RSF, mid-Nov. 1880, SLA.
38 TEF to ESH, Oct. 12, 1880, SLA.
39 RSF to SN, April 24, 1881, FBUC.
40 TEF, "log," June 11, 1881, SLA.
41 TEF on RSF, "log," Nov. 29, 1881, SLA.
42 TEF, "log," Aug. 23, 1881, SLA.
43 TEF, "log," Aug. 4, 1881, SLA.
44 TEF, "log," Sept. 4, 1880, SLA.
45 TEF, "log," July 11, 1884, SLA.
46 Ibid.
47 TEF to HEM, Sept. 23, 1882, SLA.
48 TEF, "log," Oct. 25, 1881, SLA.
49 TEF to RSF, Aug. 31, 1884, SLA.
50 RSF to ESH, May 28, 1884, SLA.
51 RSF to Grubb, Nov. 10, 1880, SLA.
52 RSF to John K. Fraser, Dec. 31, 1878, FBUC.
53 RSF to TEF, May 23, 1882, SLA.
54 RSF to W. G. McAdoo, March 3, 1879, McLC.

Chapter 6: Ladder to the sky

1 RSF to TEF, July 5, 1881, SLA.
2 RSF to ESH, June 22, 1881, SLA.
3 ESH to RSF, July 10, 1881, SLA.
4 SWB to RSF, Aug. 13, 1881, SLA.
5 HEM to W & S, Aug. 9, 1881; official letter sent Aug. 15, 1881, SLA.
6 RSF to TEF, Sept. 19, 1881, SLA.
7 Grubb to SN, May 10, 1881, SLA.
8 This account is based on C. A. Chant, "Worcester Reed Warner," *Journal of the RAS of Canada* (Jan. 1931) and Dayton C. Miller, "Ambrose Swasey, 1846–1937," *BMNAS* 22 (1943).
9 Much of this account is based on Edward Pershey's valuable thesis, *The*

Early Telescope Works of Warner and Swasey (Ann Arbor, Mich., 1982), 41–53.

10 Ibid.

11 RSF to TEF, Sept. 6, 1881, SLA.

12 Ibid.

13 ESH to RSF, July 10, 1881.

14 RSF to Sherman, Oct. 30, 1881, SLA.

15 Warner to Swasey, quoted by Pershey; Warner observed on the night of Nov. 22, 1881; RSF joined him on the night of Nov. 23.

16 W & S to TEF, June 2 and 17, 1882, SLA.

17 Ibid.

18 TEF, "log," Oct. 19, 20, 1881, SLA.

19 Staples on Kalakaua and Lick, H. H. Bancroft Collection, 45–8, BUC.

20 TEF, "log," Oct. 25, 1881, SLA.

21 TEF, "log," 1881, SLA.

22 HEM, "Diary," Nov. 17, 1881, SLA.

23 Ibid.

24 TEF, "log," Dec. 12, 1881.

25 SWB on ESH, 1881.

26 ESH to RSF, April 8, 1882, SLA.

27 TEF, "log," Dec. 12, 1881, SLA.

28 TEF, "log," Dec. 5, 1882, SLA.

29 Todd, "Diary," Dec. 6, 1882, SLA. Courtesy of the late Millicent Todd Bingham.

30 TEF on RSF and the Transit of Venus Commission, Dec. 1882, SLA.

31 Ibid.

32 TEF, "log," Jan. 2, 1883, SLA.

33 TEF, "log," Summer 1883, SLA.

34 TEF to HEM, July 14, 1883; Holden led the observatory party. A search for the hypothetical planet Vulcan was unsuccessful, but valuable spectroscopic and polariscopic observations of the corona were obtained.

35 TEF to HEM, June 9, 1883, SLA.

36 Graham arrived May 15, 1882, SLA.

37 G. W. James, *How We Climb to the Stars and the Lick Observatory* (San Francisco, 1887).

38 RSF to ESH, Dec. 24, 1883, SLA.

39 RSF to SN, 1883.

40 TEF to RSF, June 19, 1884, FBUC.

41 TEF in answer to suit brought by ESH.

42 Pershey, *Warner and Swasey.*

43 TEF at cornerstone laying, July 8, 1883, SLA.

44 Ibid.

45 SN, Report to the Lick trustees; *Harper's Magazine,* Feb. 1885.

46 RSF to ESH, Jan. 9, 1884, SLA.
47 Ibid.
48 George Clark to Alvan Clark, who reported to RSF, Dec. 27, 1884, SLA.
49 C. Feil to RSF, Oct. 28, 1884, SLA.
50 RSF to ESH, Dec. 18, 1884, SLA.
51 RSF to Clark, July 5, 1884, SLA.
52 RSF to TEF, Dec. 26, 1884, SLA.
53 TEF to RSF, Dec. 27, 1884, SLA.
54 RSF to TEF, Dec. 30, 1884, SLA.
55 RSF to TEF, Dec. 27, 1884, SLA.
56 ESH to RSF, May 12, 1882, SLA.
57 RSF to Sherman, Oct. 7, 1883, SLA.
58 RSF to D. P. Todd, Sept. 14, 1883, SLA.

Chapter 7: Success and conflict

1 TEF to RSF, March 5, 1885, SLA.
2 TEF to RSF, March 7, 1885, SLA.
3 TEF arrived there on March 7, 1885, SLA; TEF to RSF, March 17, 1885, SLA.
4 RSF to ESH, June 14, 1886, SLA.
5 TEF to RSF, April 13, 1885, SLA.
6 On April 21, 1885, RSF told ESH about Francis. On June 8, after visiting Francis and finding him interested in the Lick Observatory, ESH followed his visit with a long letter. ESH to Francis, June 8, 1885, SLA.
7 TEF to RSF, April 14, 1885, SLA.
8 TEF to RSF, April 28, 1885, SLA.
9 TEF to RSF, April 13, 1885, SLA.
10 RSF to SN, June 15, 1885, SLA.
11 RSF to ESH, Dec. 24, 1885, SLA.
12 RSF to ESH, Oct. 26, 1885, SLA.
13 RSF to ESH, Aug. 6, 1885, SLA.
14 Ibid.
15 SN to the Clarks, Nov. 4, 1885, SLA.
16 JAB to RSF, Oct. 1885.
17 SN to RSF, Oct. 29, 1886; A. Hall to SN, Nov. 19, 1886, SLA.
18 SN, *Reminiscences.*
19 TEF, "log," 1886, SLA; *SM* 6(1887):87.
20 TEF had received a telegram from the younger Feil through the Clarks, giving the same news; Sept. 4, 1885.
21 RSF to SWB, 1886, SLA.

22 TEF, "log," Dec. 26, 1886, SLA.
23 Hager to ESH, June 18, 1885, SLA.
24 RSF to ESH, Nov. 3, 1885, SLA.
25 RSF to ESH, Feb. 3, 1886, SLA.
26 ESH to Hager, July 1, 1885, SLA.
27 ESH to RSF, Jan. 20, 1885, SLA.
28 RSF to ESH, Nov. 3, 1885, SLA.
29 RSF to ESH, Feb. 3, 1886, SLA.
30 SN to ESH, Sept. 28, 1885, NLC.
31 RSF to ESH, Feb. 3, 1886, SLA.

Chapter 8: Engineering feat on Mount Hamilton

1 ESH to RSF, Oct. 24, 1884, SLA.
2 W & S to RSF, Dec. 26, 1883, SLA; Pershey, *Warner and Swasey,* 125.
3 W & S to RSF, Dec. 26, 1883, SLA; ESH, *SM* 7 (1888):57–61.
4 ESH to RSF, Oct. 25, 1884, SLA.
5 SWB to RSF, April 30, 1885, SLA.
6 ESH to W & S, May 11, 1885, SLA.
7 TEF to RSF, May 6, 1885, SLA.
8 W & S to RSF, April 9, 1886, SLA.
9 RSF to SN, May 27, 1883, SLA.
10 SN to ESH, July 12, 1883, NLC.
11 SN, *Harper's Magazine,* Feb. 1885.
12 HEM for the trustees, Feb. 2, 1886, SLA.
13 Grubb to Lick trustees, SLA.
14 Meeting of the RAS in June 1886.
15 Comment by the president, recorded in *Observatory,* June 1886.
16 Ibid.
17 HEM to Grubb, July 8, 1886, SLA.
18 The method of competitive design proposed by Newcomb would permit the trustees to choose the best of all possible designs. "Should the plan which they consider best be in itself entirely satisfactory, both with respect to mechanism and design, the mounting will be awarded to the proposer of that plan. Should modifications be necessary to the plan they will reserve the right of proposing any devices which heretofore have been used on any telescope whatever or may have been proposed or described in print by any person whatever, excepting, of course, such as are patented." Moreover, Newcomb proposed that the trustees retain the right to buy any submitted designs and ideas not actually incorporated into the telescope.
19 RSF to Grubb, Feb. 2, March 1, 1886.

20 RSF to W & S, Aug. 18, 1886, SLA.
21 Pershey, *Warner and Swasey,* 146.
22 Warner to ESH, Aug. 2, 1887, SLA.
23 SWB to ESH, Aug. 24, 1887, SLA.
24 See correspondence between RSF and W & S, 1886–7, SLA.
25 RSF to W & S, June 30, 1887, SLA.
26 W & S to ESH, April 1, 1887, SLA.
27 Holden had apparently negotiated the contract with Warner in May 1886 in San Francisco when RSF may have been on Mount Hamilton.
28 TEF to RSF, Sept. 26, 1885, SLA.
29 RSF to Plum, June 11, 1887, SLA.
30 RSF to ESH, Feb. 3, 1886, SLA.
31 RSF to SN, May 27, 1883, SLA.
32 RSF to ESH, Aug. 8, 1885, SLA.
33 TEF to RSF, Sept. 26, 1885, SLA.
34 RSF to ESH, Aug. 6, 1885, SLA.
35 TEF to RSF, Sept. 26, 1885, SLA.
36 RSF to ESH, Aug. 6, 1885, SLA.
37 Ibid.
38 *San Jose Daily Mercury,* "History," 1888. Typed version, SLA.
39 RSF to ESH, Oct. 26, 1885, SLA.
40 RSF to SN, Dec. 24, 1887, SLA.
41 George W. Dickie had joined the Union Iron Works in 1883 as manager. From 1883 to 1903 James Dickie was superintendent; *San Jose Daily Mercury,* June 28, 1888, SLA.
42 The contract was for $56,850, SLA.
43 RSF to ESH, Nov. 6, 1885, SLA.
44 Ibid.
45 HEM to the editors of *Observatory,* July 31, 1886.
46 Editors of *Observatory,* July 31, 1886.
47 Tison to M. Hamilton, Aug. 30, 1886, SLA.
48 RSF to HEM, March 21, 1888, SLA.
49 RSF to ESH, April 21, 1885, SLA.
50 RSF to HEM, Nov. 30, 1887, SLA.
51 RSF to HEM, June 17, 1886, SLA.
52 TEF, "log," 1885–6, SLA.
53 TEF, "log," June 15, 1886, SLA.
54 TEF, "log," June 16, 1886, SLA.
55 ESH, newspaper article, 1886, SLA.
56 TEF, "log," April 25, 1886, SLA.
57 RSF to ESH, April 21, 1886, SLA.
58 RSF to JEK, July 2, 1887, SLA.
59 William Keeler; see Robert W. Daly, *The Letters of Acting Paymaster*

William Frederick Keeler to his wife Anna Keeler (Annapolis, 1968); WFK Journals, 1878–86, SLA.

60 JEK, "diary," July 19–24, 1878, SLA.
61 TEF, "log," Sept. 27, 1886, SLA.
62 TEF, "log," Oct. 19, 1886, SLA.

Chapter 9: James Lick's last journey

1 Lick burial reported by *San Jose Mercury,* Jan. 19, 1887; *San Francisco Examiner.*
2 *San Jose Mercury,* Jan. 19, 1887.
3 TEF, "log," Jan. 10, 1887, SLA.
4 RSF, speech at burial of Lick.
5 RSF, rough draft of speech, 1887, SLA.
6 TEF, "log," Sept. 23, 1890, SLA.
7 Newspaper clipping, undated.
8 EEB, "A Visit to Lick Observatory," *The Santa Clara,* June 1893, 14.
9 RSF to ESH, Feb. 16, 1887, SLA.
10 RSF to ESH, Feb. 22, 1887, SLA.
11 JEK to ESH, Feb. 14, 1887, SLA.
12 RSF to J. H. Knox, San Jose, Jan. 4, 1887, SLA.
13 RSF to HEM, Jan. 26, 1887, SLA.
14 RSF to HEM, April 5, 1887, SLA.
15 TEF, "log," April 17, 1887, SLA.
16 RSF, Aug. 1887, SLA.
17 F. M. Roby to RSF, June 17, 1887, SLA.
18 *San Jose Daily Mercury,* June 28, 1888.
19 ESH, "Stellar Photography," *The Overland Monthly,* 2d ser. 11 (June 1888):587–90.
20 Clarks to RSF, May 5, 1886, SLA.
21 Clarks to RSF on death of Feil, Feb. 8, 1887, SLA.
22 Mantois to Clarks, March 7, 1887, SLA.
23 RSF to Clarks, April 7, 1887, SLA.
24 Clarks to RSF, May 1, 1887, SLA.
25 RSF to EBM, June 1, 1887.
26 Schönewald to HEM, June 10, 1887, SLA.
27 For a fuller discussion of the corrector, see John Lankford, "Photography and the Refractor," *Journal for the History of Astronomy* 14, part 2 (June 1983):77–82.
28 RSF to ESH, Sept. 2, 1887, SLA.
29 RSF to ESH, on letter from Alvan G. Clark, Aug. 13, 1887, SLA.

30 Telegram for Clark to RSF, Oct. 4, 1887, SLA.
31 Plum to RSF from Cambridgeport, May 1887.
32 Alvan Clark died on Aug. 17, 1887. A long report on Clark was published in *Scientific American,* Sept. 24, 1887; Clark's autobiography was published in SM, 8 (1889):109–17; Deborah Jean Warner, *Alvan Clark & Sons: Artists in Optics* (Washington, D.C., 1968).
33 RSF wrote to Plum in Cleveland on May 13, 1887.
34 RSF, "Diary," Oct. 12, 1887, SLA.
35 SWB to ESH, Oct. 14, 1887, SLA.
36 John A. Brashear, *The Autobiography of a Man Who Loved the Stars* (Boston and New York, 1925).
37 RSF, "Diary," Oct. 17, 1887, SLA; Harriet A. Gaul and Ruby Eisman, *John Alfred Brashear: Scientist and Humanitarian, 1840–1920* (Philadelphia, 1940), 88–9.
38 RSF, "Diary," Oct. 1887, SLA.
39 TEF, "log," Nov. 1, 1887, SLA.
40 TEF, "log," Nov. 30, 1887, SLA.
41 TEF, "log," Oct. 27, 1887, SLA.
42 TEF to RSF, Dec. 7, 1887, SLA.
43 ESH in the *New York Tribune,* Aug. 29, 1886.
44 Ibid.
45 V. A. Stadtman, *The University of California,* quoting the *San Francisco Evening Post,* June 29, 1887.
46 TEF to RSF, May 22, 1887, SLA.
47 RSF to ESH, Feb. 16, 1887, SLA.
48 RSF to ESH, draft of letter, July 18, 1887, FBUC.
49 RSF to ESH, July 21, 1887, FBUC.
50 RSF to ESH, Sept. 27, 1886, SLA.

Chapter 10: Final stages

1 RSF to ESH, Nov. 1887, SLA.
2 *San Jose Daily Herald,* Dec. 5, 1887, SLA.
3 Swasey, Letter to J. F. Holloway, July 31, 1896; reprinted in PASP 31 (1919):58.
4 Ibid.
5 RSF to HEM, Nov. 30, 1887, SLA.
6 Ibid.
7 RSF to HEM, Dec. 11, 1887, SLA.
8 RSF to SN, Dec. 5, 1887, SLA.
9 RSF to HEM, Dec. 13, 1887, SLA.

10 RSF to George Comstock, Dec. 24, 1887, SLA.
11 RSF to HEM, Dec. 24, 1887, SLA.
12 RSF, Notes, Lick Observatory, Dec. 1887, SLA.
13 Pershey, *Warner and Swasey,* 186.
14 Arrival of Alvan Clark on Mount Hamilton with his wife, Dec. 29, 1887, SLA.
15 RSF, Notes, Dec. 31, 1887, SLA.
16 Swasey, Letter, 1896 (see note 3).
17 RSF, Notes, Jan. 3, 1888, SLA.
18 Ibid.
19 Ibid.
20 RSF to HEM, Jan. 7, 1888, SLA.
21 RSF to HEM, Jan. 8, 1888, SLA.
22 RSF to HEM, Jan. 9, 1888, SLA.
23 JEK, "The First Observations of Saturn with the Great Telescope," *San Francisco Examiner,* Jan. 10, 1888, SLA.
24 RSF to HEM, Jan. 4, 1888, SLA.
25 RSF to HEM, Dec. 29, 1887, SLA.
26 Plum to RSF, Dec. 31, 1887, SLA.
27 TEF to RSF, Dec. 1887, SLA.
28 Visit from TEF, Jan. 22–5, 1888, SLA.
29 Plum on RSF, Jan. 1888, SLA.
30 RSF to SN, Feb. 18, 1888, SLA.
31 JEK, *SM* 7(1888):79.
32 Ibid.
33 Ibid.
34 *San Francisco Examiner,* Jan. 10, 1888.
35 Ibid.
36 Ibid.
37 Swasey, Letter, 1896 (see note 3).
38 Ibid.
39 RSF to SN, Feb. 11, 1888, SLA.
40 Ibid.
41 JEK to ESH, Jan. 6, 1888, SLA.
42 James Boyd to RSF, Feb. 28, 1888, SLA.
43 V. Gadesden to RSF, 1888, undated, FBUC.
44 RSF to EBM, Feb. 21, 1888, SLA.
45 RSF to SN, March 3, 1888, SLA.
46 JEK, "True Facts," *Daily Alta California,* Feb. 25, 1888, SLA.
47 Ibid.
48 Ibid.
49 John Brashear, "A Great Telescope," *Daily Alta California,* March 10, 1888, FBUC.

50 Warner to ESH, March 5, 1888, SLA.

51 Pershey, *Warner and Swasey,* discusses this problem, 192.

52 RSF to W & S, Aug. 26, 1886, quoted by Pershey, *Warner and Swasey.*

53 RSF to W & S, SLA.

54 ESH to W & S, June 23, 1888, SLA.

55 For EEB biography see R. H. Hardie, PASP Leaflet 9, nos. 415 and 416, 1964; E. B. Frost, *BMNAS,* mem. 1 21(1924).

56 EEB to RSF, 1888, SLA.

57 JAB to EEB, Sept. 1, 1887, SLA.

58 ESH, "Diary," April 4, 1888, SLA.

59 ESH to SWB, Feb. 6, 1888, SLA.

60 Ibid; Holden's complaints to regents: He complained also to Hallidie (May 11, 1888) and Phelps (Nov. 29, 1889) in a way that would have made Floyd furious. He claimed the observatory was turned over to the regents without some essential parts "just as they would be on a broken-down plantation before the war." FBUC.

61 ESH to Phoebe Apperson Hearst, Jan. 30, 1893. Phoebe A. Hearst Papers, BUC.

62 ESH to SWB, Feb. 6, 1888, SLA.

63 For more on ESH and his reign at the Lick Observatory, Donald E. Osterbrock, "Holden's Rise and Fall," *Journal for the History of Astronomy* 15(1984):81–127, 151–76.

64 Plum on RSF, 1888, SLA.

65 "Shatto was stone broke." Not long after, he died in a train wreck; Carol Green Wilson, *California Yankee* (Claremont, Calif., 1946), 40; Capt. William Banning and G. H. Banning, *Six Horses* (New York, 1930).

Chapter 11: To the stars

1 RSF to Regents of the University of California, Berkeley, April 17, 1888, SLA.

2 JEK to RSF (on visit to Mount Hamilton), April 22, 1888 (the trustees and regents arrived April 21, 1888), FBUC.

3 Ibid.

4 Charles Plum to HEM, April 30, 1888; he told of seeing Roby, Dickie, Keeler, and Holden, SLA.

5 Swasey to RSF, reported by Plum.

6 Mastick on RSF.

7 Holden's bill to the Lick Trustees, May 1, 1888, SLA.

8 SN to RSF, April 7, 1887, SLA.

9 RSF to Mastick, June 1, 1888, sent by CLF, SLA.

10 SN to RSF, July 14, 1888, SLA.
11 Trudy E. Bell, "A Labor of Love: Biography of Edward Singleton Holden, 1846–1914," *Griffith Observer* (July 1973):2–10; Orville Butler, "Edward Singleton Holden and American Astronomy, 1870–1900," M. A. thesis at the University of Notre Dame, Ind., Sept. 1981.
12 TEF to RSF, June 14, 1888, FBUC.
13 Ibid.
14 Ibid.
15 Dr. P. De Vecchi to RSF, May 23, 1888, SLA.
16 TEF to RSF, May 31, 1888, SLA.
17 TEF to ESH, June 1, 1888, SLA.
18 ESH, "Diary," June 1, 1888, SLA.
19 ESH, "Diary," June 2, 1888, SLA.
20 TEF to RSF, June 1, 1888, SLA.
21 Ibid.
22 F. M. Roby to CLF, June 14, 1888, SLA.
23 TEF to RSF, June 14, 1888, SLA.
24 Formal Recognition of the Transfer of the Lick Observatory to the Regents of the University of California, June 27, 1888, SLA. TEF to RSF, June 28, 1888, SLA.
25 Transfer.
26 Ibid; EBM at Transfer.
27 Ibid; EBM to RSF.
28 Newspaper editorial, July 28, 1888, SLA.
29 ESH to Lick trustees, Sept. 8, 1888, SLA; ESH to Phelps, Oct. 25, 1888, SLA.
30 Undated editorial, SLA.
31 Ibid.
32 CLF to SN, Oct. 16, 1888, SLA.
33 RSF to EBM, Sept. 8, 1888, SLA.
34 Ibid.
35 *LOPUB* 1(1888):13.
36 *Observatory* (Feb. 1889):120.
37 ESH to J. H. C. Bonté, Oct. 1, Nov. 9, Dec. 14, 1889, UCA.
38 EEB, *PASP* 2(1889):242.
39 See A. A. Common on Barnard's discovery, PASP 5(1887):2; EEB, *On the Photographs at the Lick Observatory in 1889,* PASP Leaflet no. 10, Sept. 1890, in Scrapbook 6, SLA.
40 JEK to ESH, Jan. 2, 1887, SLA.
41 Newspaper report on EEB, undated, SLA.
42 Bishop to RSF, April 6, 1888, SLA.
43 Hale, on honeymoon at Lick Observatory, July 10, 1890. Reminiscences in honor of Ambrose Swasey.

44 Ibid.
45 J. deBarth Shorb to RSF, Aug. 21, 1887, SLA.
46 TEF to RSF, Oct. 20, 1888, SLA.
47 *Pasadena Union,* 1889, SLA.
48 An account of this period is included in Helen Wright, *Palomar: The World's Largest Telescope* (New York, 1952).
49 JAB to RSF, June 20, 1889, FBUC.
50 An account of Floyd's 5-inch telescope was published in *SM* 15(1889):92; After RSF's death, his daughter Harry gave it to the Lick Observatory.
51 RSF to SN, Jan. 28, 1889, NLC.
52 JEK, Lick Observatory Eclipse Party. *LO Contributions* 1(1889):31; EEB, ibid., 56.
53 TEF to RSF, March 13, 1890, FBUC.
54 Ibid.
55 Ibid.
56 RSF to HALF from the Occidental Hotel, San Francisco, Aug. 11, 1890, FBUC.
57 Newspaper accounts of RSF's death and funeral ran in the *Daily Picayune* of New Orleans, the *San Jose Mercury,* and the *San Francisco Chronicle;* Holden wrote a short note, "Death of Captain Richard S. Floyd, late President of the Lick Trustees," *PASP* (1890):309–10.
58 Harkness, reported in *San Francisco Examiner,* Oct. 20, 1890.
59 SWB to ESH, quoted in *Examiner.*
60 EEB in *Examiner.*
61 JEK to W. W. Matthews, Jan. 27, 1891, SLA.
62 TEF, "Biography," *Banning Herald,* Oct. 10, 1891.
63 CLF, Obituary, *San Francisco Examiner,* Feb. 27, 1891.

Index

Made in the USA
San Bernardino, CA
30 January 2016